THE CIVIL ENGINEERING STANDARD METHOD OF MEASUREMENT IN PRACTICE

R. G. McCAFFREY FRICS
RICHARD G. McCAFFREY BTech, CEng, MICE, MIMunE
MICHAEL J. McCAFFREY DipBE, MSc, ARICS

GRANADA
London Toronto Sydney New York

Granada Publishing Limited–Technical Books Division
Frogmore, St Albans, Herts AL2 2NF
and
36 Golden Square, London W1R 4AH
515 Madison Avenue, New York, NY 10022, USA
117 York Street, Sydney, NSW 2000, Australia
100 Skyway Avenue, Rexdale, Ontario, Canada M9W 3A6
61 Beach Road, Auckland, New Zealand

British Library Cataloguing in Publication Data
McCaffrey, R. G.
The civil engineering standard method of
measurement in practice.
1. Institution of Civil Engineers. Civil
engineering standard method of measurement
I. Title   II. McCaffrey, R. G.
III. McCaffrey, M. J.
624   TA183

ISBN 0-246-11928-4

First published in Great Britain 1983 by Granada Publishing

Printed in Great Britain

Granada ®
Granada Publishing ®

Contents

Preface

The aim of this book is to provide an easy-to-use reference for those interested in the practical aspects of measurement and bill compilation in accordance with the *Civil Engineering Standard Method of Measurement* (CESMM)*. The book should be read in conjunction with the CESMM to which frequent reference is made.

Initially matters of general application are covered by a commentary on CESMM Sections 1 to 7. Subsequently, in chapters on the 24 classes of the Work Classification, the descriptive features and notes in each class are combined into Tables which summarise the CESMM rules. Commentaries on these Tables, specimen bills of quantities and measured examples are used to describe and illustrate how we apply the CESMM rules in practice. A statement of the measurement procedures and conventions adopted in the Examples is given in Chapter 1.

Our descriptions and illustrations of how we apply the rules reflect our understanding of those interpretations of the CESMM offered in *Measurement in Contract Control* by Martin Barnes (Institution of Civil Engineers - London 1977). Our experience is that these interpretations, although not official, are widely accepted in practice and we recommend *Measurement in Contract Control* to those readers who wish to study the reasoning and intentions involved in the CESMM.

As the results of the current review of the CESMM are not yet known we have not been able to include them in the book. However, we understand that there will not be any major revisions, the main aim being to clarify and amend the CESMM in order to correct any errors and remove misunderstandings.

The radical new approach of the CESMM has aroused considerable comment since its publication in 1976 and we hope, therefore, that this book will provide readers with helpful guidance on its practical application.

We thank the Institution of Civil Engineers for granting us permission to adapt the tables in the CESMM and use them in this book.

We would also like to thank Paul Rhodes for assisting with the preparation of the Drawings and are most grateful to Janet Zalega, secretary to R.G. McCaffrey, for the care she has taken in typing the script.

R.G. McCaffrey   September 1982
Richard G. McCaffrey
Michael J. McCaffrey

Bradford, West Yorkshire

* *The Civil Engineering Standard Method of Measurement, available from Telford International Bookshop, Institution of Civil Engineers, Great George Street, Westminster SW1P 3AA*

1 Measurement—Procedure and Conventions

Before proceeding with reference to the Civil Engineering Standard Method of Measurement (CESMM) it is thought appropriate to outline certain conventions and define some terms which apply to bill preparation and measurement processes generally and which are used in subsequent chapters. The present Chapter is devoted to this. Particular reference to the CESMM starts in Chapter 2.

Preparation of Bills of Quantities

Bills of Quantities are usually prepared using one of three methods, namely the traditional method, a method known as "cut and shuffle" or with the aid of a computer.

The worked Examples in this book follow the traditional method which comprises:-

1. "Taking off" - taking dimensions from drawings and entering them, with appropriate descriptions, on dimension sheets.

2. "Squaring the dimensions" - working out the areas, volumes, etc., from the dimensions and totalling the results for each item on the dimension sheets.

3. "Abstracting" - transferring the totals and items from the dimension sheets to abstract sheets in bill order and collecting together quantities for like items so that they may be cast to totals in readiness for billing.

4. "Billing" - writing the draft bill from the abstract.

Stages (2), (3) and (4) above are collectively known as "working up". Every calculation and every transferred entry performed during any stage is checked by a second person to ensure there has been no arithmetical or copying errors.

Dimension Sheets

Traditional dimension sheets are ruled as shown in Plate 1. Each sheet is ruled with an identical set of four columns each side of a vertical centre line.

The individual columns in each set of four are termed:-

| Timesing Column | Dimension Column | Squaring Column | Description Column |

Dimension Sheets (cont.

The dimension column is used for entering the dimensions and the timesing column is used for entering any multiplication factor which applies to the dimensions. The squaring column is used for entering resultants calculated from the dimensions and their multiplication factors. The description column is used for entering the descriptions of the work to which the dimensions relate. Additionally,the right hand side of the description column is used to build up dimensions; these preliminary calculations are known as "wastes". Location references of the work are sometimes entered on the right hand side of the description column.

Dimension sheets are used working from top to bottom of the left hand set of columns and then continuing down from top to bottom of the right hand set of columns. In the worked Examples in this book, only the left hand set of columns are used for dimensions. The right hand set being reserved for commentary.

Identification and Numbering of Dimension Sheets

Each dimension sheet should be headed with its job name, number or other code.

Completed dimension sheets should be numbered. On small jobs where one person is taking the whole of the dimensions they can be numbered consecutively as the measuring proceeds. On larger jobs where a number of people are engaged on taking dimensions, it is prudent to number sheets consecutively as batches are issued for taking off and to keep a running total of the sheets issued. When the completed sheets are collected a check can then be made to ensure that all sheets are accounted for. Spoiled and unused sheets should be retained until this check is complete. If necessary the collected dimension sheets should be re-arranged into a convenient order for future reference during admeasurement and re-numbered consecutively before abstracting starts. On jobs with a large number of sheets an index to the dimensions is useful.

Entering Dimensions on the Dimension Sheets

The dimensions for lengths, areas and volumes are entered in the dimension column to two places of decimals.

Drawings will usually be dimensioned in millimetres or in metres to three decimal places. Waste collections should be made using the dimensions given on the drawings. Rounding off to two places of decimals for transfer to the dimension column should be left until the waste is collected to a total.

Dimensions are set down under each other in the dimension column with a line drawn beneath, as shown on Plate 1. A single dimension with a line beneath indicates that it is a linear dimension. Two dimensions with a line beneath indicates a set of two dimensions from which the area of a surface is to be calculated. Three dimensions with a line beneath indicates a set of dimensions from which the volume of a solid is to be calculated. Numbers for numbered items are entered in the centre of the dimension column with a line beneath.

Within each set of dimensions, the dimensions are set down in the order of (1) length, (2) width and (3) depth or height. This order should always be used.

"Timesing"

Timesing is effected by entering the required number followed by an oblique

Entering Dimensions on the Dimension Sheets

	3.00	linear dimension.
	4.00 3.00	dimensions to calculate area.
	6.00 5.00 3.00	dimensions to calculate volume.

"Timesing"

3/	4	indicates multiply by 3.
3/2/	7.00 3.00	indicates multiply by 6.
2/3/2/	5.00 3.00 2.00	indicates multiply by 12.

"Dotting On"

4 · 2/	5.00	indicates multiply by (4+2) = 6.
2/3 · 2/	4.00 3.00	indicates multiply by 10.
3/·2/6/	3.00 3.00 2.00	indicates multiply by 30.

Irregular Figures

½/	6.00 3.00	area of a triangle base 6.00m height 3.00m.
22/7	5.00 5.00	area of a circle 5.00m. radius.
22/7	10.00	circumference of a circle 5.00 m radius (10.00m diam.)

PLATE 1

4

				Setting Out Items On Dimension Sheets

NOTES FOR GUIDANCE IN SETTING OUT ITEMS AND DIMENSIONS

1) BRICK RETAINING WALL - DRAWING .32/1A

2)
Wall above d.p.c.

3) <u>Engin. bkw. Class B engin. bks. as B.S.3921 in ct. mor. unless o/w. described</u>

4)
```
                    6790
         2/520= 1040
                    5750
```

2/	5.75 2.20 2.25	12.65	] Mass bkw. thickn. 520mm in vert. walls, English bond
5)	1.95	8.78	U331 (returns
		21.43	
6)	1.30 1.25	1.63	<u>Ddt</u> (want
		19.80	
7)	0.95 1.50	1.43	<u>Ddt</u> Ditto. do. (panel U331

8)
&

Add One bk. constrn., in vert. walls, Flemish bond
U311

9)

10)	5.57 5.75	NIL 5.75	] Surf. features. B.o.edge copg.
	2/ 2.25	4.50	U371
		10.25	

(1)

NOTES FOR GUIDANCE IN SETTING OUT ITEMS AND DIMENSIONS

1) Title of job or title of part of job should be entered at the head of each sheet. Alternatively a mnemonic code may be entered.

2) Side headings should be entered to indicate which particular part of the piece of work is being measured.

3) Where a particular specification type is predominant a note or heading as indicated will avoid the need to keep repeating the type in the item descriptions.

4) Note that waste calculations are entered preceding the item to which they relate.

5) Location notes as shown in the waste column are of great help in following the dimensions later. Such notes should be used extensively to identify dimensions.

6) Ddt. = Abbreviation for Deduct. Deductions which relate to immediately preceding items may be entered as shown; the deduction being made on the dimensions sheet as shown. This type of deduction will usually occur where a figure or solid with indents or voids has been measured overall and the indents or voids (known as "wants) need to be deducted.

7) Indicates a deduction which will be carried to the abstract.

8) The sign "&" is used to denote that the same dimension applies to each item description.

9) Items should be coded as CESMM.

10) Incorrect dimensions are cancelled by writing "NIL" against them in the squaring column.

PLATE 2

"Timesing" (cont.

stroke. The convention 2/ means multiply by 2. The convention 2/2 means
2 x 2 and so on. The conventions are illustrated on Plate 1.

Decimal fractions are not used in the timesing column because the decimal
point may be mistaken for the "dotting on" symbol (See next paragraph). They
are expressed as equivalent vulgar fractions. The line between the numerator
and denominator of the vulgar fraction should be horizontal, to prevent it
being mistaken for the oblique timesing stroke.

"Dotting on"

When repeating dimensions which have been timesed previously it may be found
that they cannot be multiplied and it is required to add. In these circumstances
resort is made to what is called "dotting on". A dot in the timesing column is
used in place of the conventional plus sign to indicate an addition to the
multiplying factor previously entered. The convention is illustrated on Plate 1.

Setting out Dimensions and Other Conventions

Dimensions and related descriptions should be set down on the dimension sheets
in a manner which can be easily followed and related to the drawings at a later
date. Locational references given at the right hand side of the description
column should be used wherever possible.

Deductions

When surfaces or solids are perforated by openings or their outline is indented,
it is convenient to first over measure by assuming either that they are unper-
forated or that they are of regular outline, as the case may be, and to follow
with deductions to adjust for the openings or indents. In such cases the symbol
"Ddt" is added as a prefix to the item description attached to the dimensions
which are to be deducted. See worked Examples.

Bracketing Dimensions

When more than one set of dimensions relate to the same item description, the
dimensions are bracketed. See worked Examples.

Use of Ampersand

When the same dimensions apply to more than one item description, ampersand
inserted centrally in the description column before each succeeding description
after the first, signifies that the same dimensions apply to each of the items.
See worked Examples.

Abbreviations

Abbreviations are used when writing item descriptions, during taking off and
abstracting.

There is no standard dictionary of abbreviations for civil engineering work.
It is simply a matter of adopting those which are in general use in the organisa-
tion in which one works.

In this book the abbreviations have not been reduced to the shortest

Abbreviations (cont.

practical level.   To avoid having to look up the meaning of abbreviations when examining the Examples, the aim has been to restrict them to a level readily understandable to anyone familiar with civil engineering work.

Abstracting

The sequence of items in the taking off is not necessarily the same as that of the finished bill. An abstract provides the facility to arrange items in bill order and to collect together the quantities of like items. It will be found convenient in some instances to transfer items direct from the dimension sheets to the appropriate place in the bill. This operation is termed "billing direct". When an abstract is used, it will note items which are not abstracted and which are to be billed direct. The dimension sheets will also be marked to indicate the items or series of items noted in the abstract for billing direct.

Specimen Abstract, prepared from Measured Example EE.1 in Chapter 7

C/Excavn. of cutts., soil for re-use, Excavd. Surf, 150 mm below Orig. Surf. E210.1	C/Ditto. soil for disposal do. E220.1		C/Filling. and compact to embankments sel. excavd. matl. E624		S/Fillg. ancills. trimming of slopes, nat. matl. E711
60.53 (5 37.80 (6 33.54 (7 __131.87__	230.35 (1 131.87 __98.48__	DDT 60.53 (5 37.80 (6 33.54 (7 __131.87__	986.23 (3 136.49 __849.74__	DDT __136.49__ (3	223.57 (7
			S/Fillg. and compact. thickn. 150 mm excavd. soil E631.1		S/Fillg. ancills. prepn. of surfs. nat. matl.E721.1
C/Excavn. of cutts.,matl, for re-use, Comm. Surf. u/s topsoil, Excvd. Surf. 450 mm above Fin. Surf. E230.1	C/Ditto. matl. for disposal do. E240.1		403.50 (5		279.90 (4
			S/Ditto. surfs. over 10 degrees E631.2		S/Ditto. vertical E721.2
986.23 (3 DDT 136.49 (3 __849.74__	1119.60 (2 136.49 (3 1256.09 986.23 __269.86__	DDT 986.23 (3	251.97 (6 223.57 (7 __475.54__		27.99 (4
		S/Excavn. ancills. trimming of slopes. nat. matl. E511	S/Landscaping. grass seedg., surfs. n.e. 10 degrees, Spec. Clause X E831.1		
		251.97 (6	403.50 (5		S/Ditto. over 10 degrees do. E832.1
		S/Excavn ancills, prepn. of surfs. nat. matl.E521.1			
C/Excavan of cutts. matl. for disposal, Comm.Surf. 450 mm above Fin. Surf. E240.2		440.10 (4			223.57 (7 251.97 (6 __475.54__
		S/Ditto vertical E521.2			
187.52 (2		44.01 (4			

2 The Sections of the CESMM

The rules and provisions of the Civil Engineering Standard Method of Measurement (CESMM), are applicable to the preparation of bills of quantities for and the measurement of civil engineering work. They are not intended to apply to the preparation of bills of quantities for or the measurement of mechanical engineering, electrical engineering or building work.

The CESMM is intended for use in conjunction with "the I.C.E. Conditions of Contract"* and makes reference to certain clauses in the Fifth Edition of the Conditions. The January 1979 revision of the I.C.E. Conditions of Contract provides, at Clause 57, that unless another method of measurement is stated in the Appendix to the Form of Tender, the Bill of Quantities shall be deemed to have been prepared in accordance with the CESMM.

The CESMM document contains eight numbered Sections. The first seven Sections are each divided into Paragraphs. Section 8, "Work Classification", is divided into twenty four main classifications, Termed "Classes". A classification table with Notes appended is provided for each Class. The rules, recommendations and options set out in the Notes given for a Class are particular to that Class. Those given in the Paragraphs of Section 1 to 7 are of general application. Rules, which must be observed if the provisions of the CESMM are not to be infringed, are distinguished by the use of the word "shall" in the text of any requirement expressed in a Note or Paragraph of the document. Statements in the Notes or Paragraphs which use the words "should" or "may" denote they are recommendations or options. The provisions of the CESMM will not be infringed if they are not followed. The document is consistent in the use of the words "given" and "inserted". "Given" when used in relation to the Bill of Quantities means given in the Bill before it is issued to tender. "Inserted" means inserted by the tenderers after the Bill has been issued to tender.

Regard must be given to the general matters set out in the Paragraphs when applying the Work Classification. In the present Chapter the gist of each Paragraph in the Sections of the CESMM is given, with added Commentary, as an essential preliminary to the discussion of individual Classes in subsequent Chapters. Reference should be made to CESMM for the printed text of the Sections.

---

* *"Conditions of Contract for use in connection with Works of Civil Engineering Construction" issued and approved by the Institution of Civil Engineers and the Federation of Civil Engineering Contractors and also the Association of Consulting Engineers.*

COMMENTARY

The meaning assigned to a word or expression in Section 1, will apply throughout the CESMM, except where the context otherwise requires. It will apply, also, whenever the word or expression is used in the Bill of Quantities, unless the Bill assigns it a different meaning and includes Preamble amending the CESMM accordingly.

1.1

'Conditions of Contract', means the document defined in the paragraph referenced, commonly known as the I.C.E. Contract (fifth edition).

1.2

Words and expressions which are defined in the Conditions of Contract are to be taken as having the same meaning when they are used in the CESMM.

1.3

Words and expressions which are expressed with initial capital letters when they are defined in the CESMM and the Conditions of Contract should be expressed with initial capital letters when they are used in the Bill of Quantities.

References to clauses in the CESMM, "are references to clauses numbered in the Conditions of Contract".

1.4

References to paragraphs in the CESMM, are references to paragraphs numbered in Sections 1 to 7 of the CESMM.

1.4

The definition of the word "work" as given in the paragraph referenced may be said to include all the Contractor has to do. It should not be confused with the word "Works" as defined in the Conditions of Contract.

1.5

'Expressly required'. The CESMM provides that certain work is subject to measurement only when expressly required. Such work will not qualify to be measured unless it is specifically prescribed in the Contract or specifically ordered by the Engineer.

1.6

'Bill of Quantities'. The Bill of Quantities is defined as "a list of items giving brief identifying descriptions and estimated quantities . . ."

1.7

Item descriptions in the Bill are not required to fully describe the work, they are required to clearly identify it so that its nature and extent can be ascertained from the drawings and specification. (See Paragraph 5.11 of the CESMM).

'Daywork'.   The meaning assigned to this expression in the CESMM is that generally accepted.

1.8

Provision for daywork, if required, is made in the Bill of Quantities in accordance with Paragraphs 5.2, 5.6 and 5.7 of the CESMM.

'Work Classification'. The Work Classification used in the CESMM is that set out in Section 8 of the document. It is explained in Paragraph 3.1 of the CESMM. What it defines is stated in Paragraph 2.6 of the CESMM.

1.9

COMMENTARY

The CESMM provides for what might be described as the starting
and finishing surfaces in excavation and boring items to be
identifiable from the item descriptions. (See Paragraph 5.21
of the CESMM and Commentary on that Paragraph). To simplify
the application of the requirement, the CESMM adopts a nomen-
clature for four surfaces.

'Original Surface'. This is the surface of the ground of the
Site as it exists when the Contractor enters upon it to carry
out the work of the Contract.   1.10

'Final Surface'.   This is the surface at which excavation and
boring is to finally finish as shown on the Drawings. Work
below this Surface such as soft spots, is described as 'below
the Final Surface'.   1.11

'Commencing Surface'. This is defined in relation to an item
in the Bill of Quantities, as 'the surface of the ground before
any work covered by the item has been carried out'.   The
Commencing Surface may be the Original Surface or it may be
another surface somewhere between the Original Surface and the
Final Surface.   1.12

'Excavated Surface'.   This is defined in relation to an item
in the Bill of Quantities, as 'the surface to which excavation
included in the work covered by the item is carried out'. The
Excavated Surface may be the Final Surface or it may be another surface somewhere between the Final Surface and the
Original Surface.   1.13

    An application of the definitions of the Surfaces is
illustrated in Figure E3 in Chapter 7.

'A hyphen between two dimensions'. This describes the abbreviation used to denote a range of dimensions which includes all
dimensions exceeding the first dimension but not exceeding the
second.   1.14

GENERAL PRINCIPLES

The general Principles set out in Section 2 of the CESMM, provide
some rules but they are mainly general statements.

Title, Application and Extent

The title of the document is confirmed as the "Civil Engineering
Standard Method of Measurement". The title may be abbreviated to
"CESMM". It is intended for use in conjunction with the I.C.E.
Conditions of Contract (fifth edition) in connection with civil
engineering work.   2.1

    Work other than civil engineering construction, for which
the CESMM is not appropriate, shall be itemised in sufficient
detail to enable tenderers to price it adequately. Where such
work is measured, the method of measurement shall be stated in
the Preamble to the Bill of Quantities in accordance with Paragraph 5.4 of the CESMM.   2.2

COMMENTARY

Title, Application and Extent (cont.

    Work for which the CESMM is not appropriate, which is
itemised in the Bill of Quantities otherwise than provided
in the CESMM, will be deemed to be "itemised in sufficient
detail" if the Contractor prices it without question. When
the Contract is placed, the Contractor will have no grounds
to sustain a claim that such work is itemised in insufficient detail, if he has priced the work without
qualification.

Object of the CESMM

    In defining the object of the CESMM, the Paragraph refer-
enced provides the rules that the Bill of Quantities shall
be prepared and priced and the quantities of work shall be
expressed and measured all in accordance with the CESMM.   2.3

    It is permissible to deviate from the above rules
subject to the provisions of Paragraph 5.4 of the CESMM
being observed. Such deviation should be limited to those
items or sections of the work for which the CESMM would be
inappropriate.

Objects of the Bill of Quantities

The Bill of Quantities provides information of quantities
to enable tenders to be prepared and provides a means of
valuing work after the Contract is placed, all as stated
in the Paragraph referenced.   2.4

    Items in the Bill of Quantities should "distinguish
different classes of work". This is achieved by itemising
the work in accordance with the CESMM. For work of the
"same nature carried out in different locations or in any
other circumstances", the person compiling the Bill should
adjudge the cost characteristics of location or circumstances
and provide items in the Bill of Quantities which separate
different considerations of cost.   2.5

    The CESMM suggests that consistent with the foregoing
itemisation requirements, the Bill of Quantities should be
as simple and brief as possible.   2.5

    The rules which Work Classification provides for the
purpose of preparing quantities for work, govern the division
of work into items, the information to be given in item
descriptions, the units for the quantities and the measure-
ment of work to calculate the quantities.   2.6

Work Classification is the system introduced by the CESMM for
the purpose of preparing and presenting the quantities of work.
It governs the division of work into items, the information to
be given in item descriptions, the units for the quantities and
the measurement of work to calculate the quantities. Its
general application is outlined in Section 3 of the document.

Item Descriptions

The Work Classification sets out groups of not more than eight
phrases, termed "descriptive features", in each of three
"divisions" to create a classification table for each of the
twenty four Classes into which the Work Classification Section
of the CESMM is divided. An outline of what is included and/or
excluded follows the Class title. Notes are appended following
the classification table of each Class.   3.1

The Class appropriate to a component of work is selected
by reference to the title of the Class and to the "Includes"
and the "Excludes" given preceding the classification table.

Item descriptions are compiled by combining listed
descriptive features. Each item description is made up of one
descriptive feature from each division taken from between the
same pair of horizontal lines in the classification table of
the appropriate Class. In some cases these descriptions need
to be amplified by additional description. See "Additional
Description", Paragraph 3.2 of the CESMM and subsequent
Commentary.

Item descriptions need not use precisely the words of the
Work Classification. It is, nevertheless, sensible to do so
except where it would result in the duplication of information.
It is unnecessary to include standard descriptive features in
item descriptions which give the same information in greater
detail in accordance with any Note or provision of the CESMM.
For example, the item description "Douglas fir decking, thickness 63 mm", would be adequate without stating that the decking
is softwood and that it is in the thickness range 50 - 75 mm
(Refer to Class O of the CESMM - First division feature 6,
Second division feature 3 and Note O2).

Additional Description

Item description compiled by taking one descriptive feature
from each division, as provided in Paragraph 3.1 of the CESMM,
must be amplified by any additional description required by
any provision of Section 5 of the CESMM or any applicable Note
in the CESMM Work Classification.   3.2

COMMENTARY

Mode of Description

Item descriptions given in the Bills of Quantities for Permanent
Works are required by the CESMM to identify the component of
the Works and not the tasks to be carried out. The descriptive
features in the Work Classification are phrased accordingly and
their use in item descriptions will ensure compliance with the
Paragraph referenced.   3.3

    Item descriptions for work specifically limited must state
the limitation. To conform with this provision, additional
description in the form of an ad-hoc qualification, identifying
the applicable limitation is needed. For example, "Fix-only",
will require to be included in the item description, or in the
heading for a group of similar items, limited to this extent.   3.4

Separate Items

An item for a component of work comprises not more than one
feature from each division of any one Class. Each item is given
separately in the Bill of Quantities. Two or more items may not
be grouped into one item.   3.5

    Separate items in the Bill of Quantities must distinguish
each difference created by any additional description attached
to otherwise similar items.   3.6

    Notes in the CESMM which point out that separate items are
or are not required for a particular component of work do not
extend or limit the itemisation required by the CESMM.   3.7

    Where there are differences in work which does not require
to be separately itemised which it is adjudged relate to a
similar item, the item should be divided to identify each
difference by separate items. For example, similar brickwork
required to be tied with non-ferrous ties would be given separately from that required to be tied with ferrous ties. (Refer to
Class U - Note U3 of the CESMM).

    The provision that separate items are not required does not
mean that there is no need for the work to be prescribed in the
Contract. Work for which separate items are not required is
included to the extent that the work is specified or otherwise
prescribed in the Contract.

Units of Measurement

The unit of measurement in which the quantities are to be given
for an item in the Bill of Quantities is that stated for the
item in the Work Classification. The unit stated against a descriptive feature in the Work Classification will apply to all
items to which the feature applies.   3.8

COMMENTARY

Refer to CESMM
Paragraph

The CESMM provides code numbers for the items in the Work
Classification. The system of coding is explained in
Section 4 of the document.

Coding

The code number for an item consists of the Class letter
and the numbers of the descriptive features in the first,
second and third divisions of the Work Classification, in
that order. For example, E421 is the code number for the
item "General excavation, topsoil for disposal, maximum
depth not exceeding 0.25 m" (Refer to Class E - Earthworks -
CESMM).   4.1

For reference purposes, the CESMM uses the symbol *
to denote all numbers in a particular division.   4.2

Item Numbers

The code numbers given in the CESMM, may be used as item
numbers in the Bill of Quantities. Whether they are so
used is optional. Bills of Quantities are usually divided
into Parts. When code numbers are used for numbering the
items in the Bill, it is convenient to list the items in
order of ascending code number within each Part.   The
suggested order of listing would not be followed where it
was considered that a series of items within a Part of the
Bill, could be more usefully grouped under a heading.   4.3

Code numbers used as item numbers in the Bill of
Quantities are not part of the item descriptions. For the
purpose of the Contract the items are interpreted on the
wording of the descriptions. An incorrect code number does
not cause a correct description to be incorrect. A correct
code number does not cause an incorrect description to be
correct.   4.4

Coding of Unclassified Items

Where an item includes a feature which the Work Classification does not list, the digit 9 is used in the applicable
division position in the code number. For example, U925
would be the code number for the item "Refractory brickwork,
Specification clause 23/10, half brick construction, facing
to other materials". (Refer to Class U - Brickwork, etc. -
CESMM).   4.5

For an item to which a division of classification does
not apply or for which the Work Classification lists less
than three divisions of classification, the consequent
absence of a descriptive feature in the item is denoted by
the use of the digit O in the applicable division in the
code number. For example, E250 is the code number for the
item "Excavation of cuttings, rock for re-use". (Refer to
Class E - Earthworks - CESMM).   4.6

COMMENTARY

Numbering of Items with Additional Description

To distinguish items which have the same code number
but different additional description a suffix number
is attached to the code number. For example, the
item "Filling and compaction, thickness 150 mm,
excavated topsoil" and a similar item of a thickness
of 225 mm, which were included in the same Part of
the Bill, would be coded E631.1 and E631.2, respectively. (Refer to Class E - Earthworks - CESMM).

4.7

PREPARATION OF THE BILL OF QUANTITIES

Section 5 of the CESMM contains rules and recommendations regarding the preparation and presentation of
the Bill of Quantities, including its format and content.

Measurement of Completed Work

The rules of measurement given in the CESMM when applied
to the preparation of the Bill of Quantities for a
Contract apply equally to the measurement of the completed
work on that Contract.

5.1

Sections of the Bill of Quantities

The Bill of Quantities is required to be divided into
the sections listed in the Paragraph referenced. The
Sections are as follows:-

    A.   List of principal quantities

    B.   Preamble

    C.   Daywork Schedule

    D.   Work items grouped into parts

    E.   Grand Summary

5.2

Each section of the Bill is usually identified by the
capital letter as indicated for the section in the list
above. Section D groups "Work Items" into Parts. Each
Part is usually numbered. (See Paragraph 5.8 of the CESMM
and subsequent Commentary). If not required Section C -
Daywork, may be omitted. (See Paragraphs 5.6 and 5.7 of
the CESMM and subsequent Commentary).

List of Principal Quantities

The list of principal quantities provided in Section A of
the Bill of Quantities will consist of a brief schedule
of the overall quantities of the adjudged principal
components of the Works. (An Example is given in Figure 1.)
It will not usually show separately the quantities for
individual Parts. It is of no contractual significance if
there are differences between the list and the detailed
quantities.

5.3

SECTION A. - LIST OF PRINCIPAL QUANTITIES

Provisional Sums	£ 55000
Prime Cost Items	£ 10000
Excavation	650 m3
Filling	200 m3
Concrete - mass in situ	110 m3
Pipelines - 975 - 1200 mm concrete pipes	750 m
Pipelines - 900 mm GRP pipes	250 m
Manholes	15 nr
Tunnels - 1500 mm concrete bolted segments	45 m
Shafts - 5280 mm concrete bolted segments	14 m
Filling old sewers	350 m

Fig. 1. Example of List of Principal Quantities for imaginary contract.

SECTION C. - DAYWORK SCHEDULE

Refer to clause 52(3) of the Conditions of Contract.

The Contractor shall be paid for work executed on a Daywork basis at the rates and prices and under the conditions contained in the "Schedules of Dayworks carried out incidental to Contract Work" issued by the Federation of Civil Engineering Contractors current at the date of the execution of the Daywork adjusted as follows.

Labour	addition/deduction*	of	....+ per cent
Materials	addition/deduction*	of	....+ per cent
Plant	addition/deduction*	of	....+ per cent

*To be deleted where inappropriate by the Contractor when tendering

+To be inserted by the Contractor when tendering

Such adjustment shall be made after the percentage additions to the amount of wages and the cost of materials contained in the aforementioned "Schedules of Dayworks" have been applied.

Fig. 2 Example of Daywork Schedule in accordance with sub-paragraph (b) of Paragraph 5.6 of the CESMM.

COMMENTARY

Preamble

Work not covered by the CESMM is sometimes required to be included in the Bill of Quantities. If this work is measured,
the Preamble must state the methods of measurement and/or any
amendments to the CESMM which have been adopted when preparing
the Bill of Quantities for the work and which are to apply when
it is re-measured. Items in the Bill for such work or headings,
where the items are grouped under headings, must make specific
reference to the applicable Preamble.   5.4

A definition of rock must be given in the Preamble where
the Bill of Quantities includes items for excavation, boring or
driving.   5.5

The terms in which rock is to be defined are not given in
the CESMM. Preferably definition gives geological description
or rock types in the terms used to describe the materials in the
borehole logs or reference conditions and classifies them according to weathering indices or mechanical properties, such as
strength, fracture spacing, bedding characteristics, etc. The
objective of the classification is to provide a means of identifying particular strata which will qualify, under the Contract,
to be measured as rock or a particular class of rock.

The function of the Preamble is not restricted to that outlined for it in Paragraphs 5.4 and 5.5 of the CESMM. Even when
no deviation is made from the CESMM, circumstances arise in the
practical application of its provisions which suggest that
certain conventions and item coverage, which might in the circumstances not otherwise be clear, should be explained in Preamble
for the guidance of tenderers.

Daywork Schedule

The CESMM provides that the Daywork Schedule, if any, included
in the Bill of Quantities shall take either of two forms, viz:-

    (a) a specially prepared list as described in sub-
        paragraph (a) of Paragraph 5.6 of the CESMM, or   5.6(a)

    (b) a statement on the lines of sub-paragraph (b) of
        Paragraph 5.6 of the CESMM, bringing into use the
        stated "Schedules of Dayworks"issued by the Federa-
        tion of Civil Engineering Contractors, and providing
        for tenderers to insert the percentage addition or
        deduction which they require to apply to the rates
        and prices in the "Schedules of Dayworks" for each
        of the stated elements of Daywork.   5.6(b)

Because of the considerable work involved in setting out
definitions and conditions for a Daywork Schedule in the form at
(a) above, the use of such a Schedule is normally restricted to
circumstances where a satisfactory means of valuing work executed
on a daywork basis could not be achieved by a Daywork Schedule
which makes use of approved published Schedules of Daywork.

The form indicated at (b) above is that which is commonly
used. In the Bill of Quantities it would be given the heading,

COMMENTARY

Daywork Schedule (cont.

"SECTION C - DAYWORK SCHEDULE", which may be followed by
a reference to clause 52(3) of the Conditions of Contract.
Thereafter, the complete text, except for the first three
words, of sub-paragraph (b) of Paragraph 5.6 of the CESMM
would be repeated. (See Example Figure 2.). The further
element of supplementary charges covered by Schedule 4
of the stated "Schedules of Dayworks" would be added where
expenditure on this element of daywork is anticipated.

Provisional sums for work executed on a Daywork
basis may be given in the Bill of Quantities. When this
is done, each element of Daywork forms the subject of a
separate item giving an associated provisional sum. The
sum being an estimate of the amount which might reasonably
be expected to be expended on the element. The items with
their associated provisional sums would be given in the
Bill of Quantities in Class A - General Items. (See
Items A411 - A416 of the CESMM). Each of these items
would be followed by an item for adjustment, if the
Daywork Schedule used is of the form indicated in (b)
previously. The item for adjustment would provide for
the appropriate percentage addition or deduction, inserted
in the Daywork Schedule by the Contractor, to be applied
to the associated provisional sum of the preceding item in
calculation of the price of the item to be included in the
amount of the Bill of Quantities.

5.7

Work Items

Division of the Bill of Quantities into Parts

It is customary to divide the Bill of Quantities into Parts.
The recommendation implied by the CESMM is that the Bill
should be arranged in numbered Parts to distinguish between
those parts of the work which for any reason is thought
likely to give rise to different methods of construction or
considerations of cost. Class A - General Items will usually
be given as a separate Part of the Bill, although it may
sometimes be considered more helpful for General Items
specifically allied to one Part to be given with that particular Part. Throughout the Bill of Quantities, items within
each Part must be arranged in the general order of the Work
Classification.

5.8

Parts are created for sections of the Works. A Part may
include several Classes. Within the Parts, the work may be
such that it requires items to be sub-divided (See Paragraph
5.10 of the CESMM).

The Parts into which the Bills of Quantities is divided
is decided by the person responsible for the preparation of
the Bill on the basis of their own judgment.

COMMENTARY

Headings and Sub-headings

A heading must be given to each Part of the Bill of
Quantities. Within each Part items may be grouped
under sub-headings. Headings and sub-headings form
part of the item descriptions to which they apply.
They must be repeated at the start of each new page
which lists items to which they apply. In the Bill
of Quantities, a line must be drawn across the item
description column below the last item to which each
heading or sub-heading applies.   5.9

The heading given to a Part of the Bill of
Quantities will indicate the section or element of
the Works covered by the particular Part.

Creating a sub-heading consisting of what otherwise would be repetitive features in several item
descriptions contributes to the brevity of the Bill
of Quantities.

An example of the use of headings and sub-
headings, also the application of the rule in Paragraph 5.9 of the CESMM which requires a line to be
drawn across the item description column,is illustrated in Figure 3.

Extent of Itemisation and Description

Work must be described and given in items as provided
in the Work Classification. Where work for any reason
is thought likely to give rise to special methods of
construction or consideration of cost, the CESMM suggests
that which causes the work to be considered special may
be identified by providing additional description and
items, beyond the requirements of the Work Classification.   5.10

The person responsible for the preparation of the
Bill of Quantities will decide, using their own judgment,
where to provide additional description and items, as
referred to in the preceding paragraph. Additional
description and items would not be provided to identify
that which would have little or no cost effect.

Descriptions

"Descriptions (in the Bill of Quantities) shall
identify the work covered by the respective items".
The extent and nature of the work is to be ascertained
from the other Tender/Contract documents read in conjunction with the Work Classification.   5.11

PART 5. WALLS - SERVICE AREAS

Number	Item description	Unit	Quantity	Rate	Amount	
					£	p
	BRICKWORK, BLOCKWORK AND MASONRY					
	Engineering brickwork in Class B bricks as Specification clause 4.3, in Type A mortar as Specification clause 4.7					
U311	One brick construction, vertical walls.	m2				
U334	Mass brickwork, thickness 328 mm, piers.	m2				
	Surface features					
U371.1	Copings, brick on edge, one brick walls.	m				
U371.2	Copings, brick on edge, 328 mm piers.	m				
U378	Fair facing, Specification clause 4.22	m2				
	Brickwork ancillaries					
	Damp proof courses, to BS 743, type E					
U382.1	Width 215 mm.	m				
U382.2	Width 328 mm.	m				
U386	Built in pipes, cross-sectional area 0.025 - 0.25 m2.	nr				

COMMENTARY

The line drawn across the description column beneath Item Code U378 denotes the end of the items to which the sub-heading "Surface features" relates, but leaves both the sub-heading "Engineering brickwork, etc." and the main heading "BRICKWORK, BLOCKWORK AND MASONRY" operative.

The line drawn across the description column beneath Item Code U382.2 denotes the end of the items to which the sub-heading "Damp proof courses etc." relates, but leaves the sub-headings "Brickwork ancillaries" and "Engineering Brickwork, etc." and the main heading "BRICK-WORK, BLOCKWORK AND MASONRY" operative.

The three lines drawn across the description column beneath Item Code U386 denotes the end of the items to which the sub-headings "Brickwork ancillaries", "Engineering brickwork, etc." and the main heading "BRICKWORK, BLOCKWORK AND MASONRY" relates.

Fig. 3.   Illustration of lines drawn across description column
as required by Paragraph 5.9 of the CESMM.

20	90	10	20	20	20	8
				Page total		

Fig. 4. Column widths in mm of Bill paper as suggested in
Paragraph 5.22 of the CESMM.

COMMENTARY

Descriptions (cont.

Subject to Paragraph 3.2 of the CESMM, item descriptions prepared in accordance with the Work Classification are sufficient provided they clearly identify the work they represent. It is appropriate to provide further itemisation and additional description where they would not otherwise clearly identify the work. See also Commentary on Paragraph 5.13 given subsequently.

An appropriate Drawing or Specification reference may be given in an item description in place of any detail of description required to be given in accordance with the Work Classification.

5.12

Additional description stating location or other physical features shown on the Drawing or described in the Specification must be given where an item description compiled in accordance with the Work Classification would be insufficient to clearly identify the work it represents.

5.13

Ranges of Dimensions

The Work Classification provides for the dimensions of some components to be given within stated ranges. The description of an item must state the actual dimension in place of the range where all the components in the item are of one dimension.

5.14

Implementing the foregoing is made simpler if when taking dimensions a note is made of the actual dimension in the description of the component concerned. Otherwise if only the range is entered when taking off it will not be apparent, without referring to the dimension sheets, whether or not several items within a range collected together on the abstract are all of one dimension. (See subsequent Measured Examples).

Prime Cost Items

The expressions "Prime Cost Item" and "Nominated Sub-contractor" when used in the CESMM, will have the meanings given in clauses 58(2) and 58(5), respectively, of the Conditions of Contract.

Prime Cost Items must be provided in the Bill of Quantities for all work which is to be carried out by Nominated Sub-contractors, whether or not the Nominated Sub-contractor is to carry out work on the Site. The Prime Cost Item given in the Bill of Quantities will consist of an identifying description with an associated sum. The sum being an estimate of the cost of the work to be carried out by the Nominated Sub-contractor. Each Prime Cost Item given in the Bill of Quantities must be followed by:-

5.15

   (a)   an item, providing for the Contractor to insert a sum for labours in connection with the Prime Cost Item, which unless the Contract expressly requires otherwise, will include only

Prime Cost Items (cont.

        the work stated in sub-paragraph (a)(i) of
Paragraph 5.15 of the CESMM when the
Nominated Sub-contractor is to carry out
work on Site, and only the work stated in sub-
paragraph (a)(ii) of the same Paragraph when
the Nominated Sub-contractor is not to carry
out work on Site, and   5.15(a)

(b) an item providing for the Contractor to insert
as a percentage of the Prime Cost Item all
other charges and profit.   5.15(b)

Where the Contractor is required to provide for the
Nominated Sub-contractor labours other than or in addition
to those indicated in Paragraph 5.15 of the CESMM, the
precise requirements must be prescribed in the Contract.
They must be stated in the descriptions of items designated
"special labours" given separately from the standard
"labours" items. See Note A10 of the CESMM.

Where the Contractor is to use any goods, materials or
services supplied by a Nominated Sub-contractor, the Bill
of Quantities will need to provide items for the Contractor's
work in using them. This could be done in the Bill of
Quantities in one of the following ways:-

(i) Insofar as the CESMM is appropriate for the work
it can be measured and billed in accordance with
the Work Classification, subject to Paragraphs
3.4 and 5.16 of the CESMM. For Paragraph 3.4 see
previous Commentary. For Paragraph 5.16 see subsequent Commentary which gives that reference.

(ii) Where the CESMM is not appropriate to the measurement of the work it can be itemised as provided in
Paragraph 2.2 of the CESMM. See previous Commentary
which gives that Paragraph reference.

(iii) The work can be made the subject of a Provisional
Sum.

The procedures as (i) to (iii) above would be appropriate
also to any Contractor's work in connection with the work
carried out by a Nominated Sub-contractor, as well as to that
which uses the materials, etc., the Nominated Sub-contractor
supplies.

The items for the Contractor's work dealt with as (i) above,
can be given in the Bill of Quantities, suitably identified in
their appropriate position in the Class to which they belong. It
is considered, however, that they are better grouped under an
identifying heading at the end of the Class to which they belong
or in a separate Part of the Bill created for the purpose. Items
for Contractor's work dealt with as (ii) above can be grouped
under an identifying heading at the end of any Class considered
allied to that to which they belong or in a separate Part of the
Bill created for the purpose. Where Provisional Sums are given

COMMENTARY

Prime Cost Items (cont.

for the Contractor's work as (iii) above, they are the
equivalent of specified contingencies and would be
given with associated identifying descriptions in the
General Items of the Bill of Quantities in the same way.

Items or headings under which are grouped items
for work for which the Contractor is to use goods or
materials or services supplied by a Nominated Sub-
contractor must make reference to the Prime Cost Item
under which they are supplied.

5.16

Provisional Sums

For definition of the term "Provisional Sum" when used
in the CESMM, see clause 58(1) of the Conditions of
Contract.

Any provision for contingencies required to be made
in the Bill of Quantities must be made by giving Provisional Sums and not by giving quantities of work in
excess of expected requirements. Provisional Sums for
any specified contingencies are given with associated
identifying descriptions in the General Items of the
Bill of Quantities. Any Provisional Sum for a general
contingency must be given in the Grand Summary. (See
Paragraph 5.25 of the CESMM and subsequent Commentary).

5.17

The CESMM outlaws the deliberate increasing of
quantities beyond the computed total requirements as a
means of providing for contingencies. The forbearance
imposed by the CESMM in this respect does not mean that
the quantities given should be the minimum quantities
and that all variables should be made the subject of
Provisional Sums.   Where there are components which
have dimensions which may vary there is no reason why
they should not be estimated so that quantities may be
computed to give the expected total requirements. If
this is done, it would be consistent with the rule
established by Paragraph 5.18 of the CESMM, which states
"quantities shall be computed ... from the Drawings" to
judge the extent of the variables and show them on the
Drawings to enable the total expected quantities to be
computed as stated. Alternatively, in any case where
variables are not shown and it is judged that more work
than is shown on the Drawings is to be expected, the
procedure would be to compute the quantities from the
Drawings and give a Provisional Sum to cover the cost
of the additional work which might reasonably be
expected.

Quantities

The quantities given in the Bill of Quantities must be
computed net from the Drawings, unless the CESMM directs
otherwise. They will usually be rounded to whole units.
Fractional quantities of 0.50 and over being rounded up
to the next whole unit. Other fractional quantities

COMMENTARY

Quantities (cont.

being neglected. Nevertheless, the use of fractional
quantities may be considered where quantities are small.
The CESMM imposes no rule regarding rounding off quantities.
It suggests that fractional quantities are not generally
necessary and should not be given to more than one decimal
place.   5.18

Units of Measurement

The Paragraph referenced sets out the abbreviations used
for the units of measurement previously referred to in
Paragraph 3.8 of the CESMM.   5.19

The abbreviations use small case letters with no full
stops. This convention must be followed when they are used
in the Bill of Quantities.

Work Affected by Bodies of Water

"Bodies of water" includes rivers, streams, canals, lakes
and such like as well as tidal water. Item descriptions
must distinguish work affected by bodies of water. They
are not required to distinguish work which is affected by
ground water whether tidal or otherwise. Item descriptions
must also distinguish between work affected at all times and
that affected only at some states of the tide where the body
of water which affects the work is tidal. Additionally, the
water surface levels, used when making the distinction referred to in the preceding sentence, must be stated in the
item descriptions.   5.20

A definition of what constitutes "work affected by bodies
of water" is not given in the CESMM. The person preparing
the Bill must exercise their judgement and distinguish any
work they consider to be affected in any way by the existence
of bodies of water. The distinction will apply to work in,
over, adjoining, abutting or in any situation in relation
to a body or bodies of water where the work is affected by
something which would not obtain if the body or bodies of
water did not exist. It will not apply to work associated
with or in conjunction with work affected by bodies of water
which is not itself affected. For example, the surfacing,
kerbing, white lining etc., of the road work carried out
after the completion of the deck and balustrade of a bridge
over water, is considered not to be "work affected by bodies
of water".

Item descriptions distinguish work affected by bodies of
water by including additional description which identifies
the body of water which affects the work. Item descriptions
for work affected by tidal water are further amplified to
indicate that the work is to be carried out at positions
affected by the identified body of water either at all times
(i.e. when the tide is at low water level) or, when the tide
is between low and high water level (i.e. at some states of
the tide). They will state also the water surface levels of
both mean low water ordinary spring tides and mean high water

COMMENTARY

Work Affected by Bodies of Water (cont.

ordinary spring tides used in making the distinction
referred to in the preceding sentence. The water
surface levels will usually be related to Ordnance Datum
(Newlyn). In the majority of cases the distinction
"work affected by bodies of water" will apply to groups
of items. In such cases headings and sub-headings may
be given which when combined provide the additional
description for the group of items to which they apply.
An example of this arrangement is given in the specimen item descriptions, Figure P1, in Chapter 12.

Ground and Excavation Levels

Item descriptions for work which involves excavation,
boring or driving must identify the Commencing Surface
when it is not also the Original Surface. They must
identify the Excavated Surface where it is not also the
Final Surface. The depths of work given in the item
descriptions in accordance with the Work Classification
must be measured from the Commencing Surface to the
Excavated Surface.

5.21

The identification required as provided in the preceding paragraph is effected by stating in the item
descriptions the position of the Commencing Surface or
the Excavated Surface, as the case may be, in relation
to the Original Surface or the Final Surface or to a
surface of a component of the Works. Descriptions do not
identify Surfaces by level. It is not necessary to state
a Commencing Surface or an Excavated Surface where these
are the Original Surface and the Final Surface, respectively. Examples of the application of the rule given in
Paragraph 5.21 are given in measured Examples in Chapter
7.

Form and Setting Out

The CESMM suggests the Bill of Quantities is reproduced
on paper of A4 size. It suggests also how paper on which
to set out work items should be ruled and headed. The
ruling and headings are shown in Figures 3 and 4, with
column widths in millimetres superimposed on the suggested
sheet.

5.22

Each page of the Bill of Quantities makes provision
for the amounts inserted to be brought to a total. See
Figure 4. A Summary must be provided for each Part of the
Bill of Quantities. It must make provision which enables
the amount of each page to be brought into the Summary and
cast to a total. The Summary will indicate also that the
total amount of the Part is to be carried to the Grand
Summary. (See Paragraph 5.24 of the CESMM). Paper ruled
differently from that for the work items will usually be
used for Summaries. An example of a Part Summary is shown
in Figure 5.

5.23

```
PART 1 - GENERAL ITEMS - SUMMARY         £    p

          Page 11 .  .  .

          Page 12 .  .  .

          Page 13 .  .  .

          Page 14 .  .  .


     TOTAL PART 1 - Carried to
     SECTION 'E' GRAND SUMMARY  £
```

Fig. 5.   Example of a Part Summary prepared to comply with
          Paragraph 5.23 of the CESMM.

SEA DEFENCE WORK AT TINEMOUTHEND


```
SECTION E - GRAND SUMMARY                        £       p

   PART 1  General Items .  .  .  .  . Page 15 .  .

   PART 2  Site Clearance   .  .  .  . Page 19 .  .

   PART 3  Sea Wall   .  .  .  .  .  . Page 28 .  .

   PART 4  Promenade Pavements .  .  . Page 32 .  .

   PART 5  Flood Walls   .  .  .  .  . Page 39 .  .

   PART 6  Work to Outfall Sewer  .  . Page 46 .  .

                              Total   £

General Contingency Allowance                  25000    00

                              Total   £

Adjustment Item : Addition/Deduction*  .  .  .  .

                       Tender Total    £
```

Signed . . . . on behalf of . . . .

Date   . . . .

        * Delete as appropriate

Fig. 6. Example of a Grand Summary prepared to comply with
         Paragraphs 5.24 to 5.27 (inclusive) of the CESMM

COMMENTARY

	Refer to CESMM Paragraph

Grand Summary

The last section of the Bill of Quantities gives a Grand Summary, (See paragraph 5.2 of the CESMM). The Grand Summary must give a list of Parts and provide a column in which the amounts from the Part Summaries can be inserted and totalled.

5.24

General Contingency Allowance

Any provision in the Bill of Quantities for a general contingency must be made the subject of a provisional sum. (See the last sentence of paragraph 5.17 of the CESMM). The sum must be given in the Grand Summary in a position immediately following that provided for the total of the amounts for the Parts.

5.25

Adjustment Item

An Adjustment Item must be given in the Grand Summary. It must be the last item given in the Grand Summary. Provision shall be made for all amounts other than that for the Adjustment Item to be brought to a total immediately preceding the Adjustment Item. (See also Paragraphs 6.3 and 6.4 of the CESMM).

5.26

Total of the Priced Bill of Quantities

The Grand Summary must make provision for all amounts in the Summary to be brought to a Grand Total. (See Figure 6).

5.27

COMPLETION AND PRICING OF THE BILL OF QUANTITIES

Rules and explanatory statements in relation to the completion and pricing of the Bill of Quantities by a Tenderer are set out in Section 6 of the CESMM.

Insertion of Rates and Prices

Rates and prices are required to be entered in the rate column by the tenderer in pounds sterling with pence entered as decimal fractions of one pound.

6.1

Specific reference to the rate column in Paragraph 6.1 of the CESMM, indicates the intention that tenderers must insert a rate for each item intended to be priced. A lump sum inserted in the amount column for a group of bracketed items is not acceptable.

When the ICE Conditions of Contract (fifth edition) are used, items which a tenderer chooses to leave unpriced are deemed to be covered by the priced items. See Clauses 11(2) and 55(2) of the ICE Conditions of Contract.

COMMENTARY

Parts to be Totalled

Tenderers are required to cast the extended prices to a
total for each Part of the Bill of Quantities and to
carry the total of each Part to the Grand Summary. This
is facilitated by giving a summary for each Part (See
Paragraph 5.23 of the CESMM).   6.2

Adjustment Item

A tenderer may insert a lump sum addition or deduction
against the Adjustment Item (given in the Grand Summary)
in adjustment of the total of the Bill of Quantities.   6.3

Any amount inserted by the Contractor in respect
of the Adjustment Item constitutes a fixed lump sum
adjustment. It will be treated as such in the final
settlement whether or not the total value of work is
more or less than that anticipated in the Bill of
Quantities. It will, however, be subject to adjustment in accordance with the Baxter Formula*where the
Contract includes provision for adjustment in respect
of cost fluctuations by the use of that Formula.

Instalments on account of the amount, if any, of
the Adjustment Item must be made in interim certificates.
They are made in the proportion that the amount referred
to in clause 60(2)(a) of the Conditions of Contract bears
to the total of the Bill of Quantities before the
addition or deduction of the amount of the Adjustment
Item. They are made before the deduction of retention.
They must not exceed in aggregate the amount inserted
for the Item in the Bill. When for an interim certificate
the calculations by proportion show that the aggregate of
the instalments has reached or would exceed the amount
originally inserted, the amount certified in that certificate and which would continue to be certified in
subsequent certificates would be the amount of the
Adjustment Item. When at the date of the issue of the
Certificate of Completion for the whole of the Works (See
clause 48 of the Conditions of Contract) the aggregate
of the instalments has not reached the amount of the
Adjustment Item, the balance must be added or deducted,
as the case may be, in the next certificate after the
issue of the Certificate of Completion.   6.4

Loosely defined the amount referred to in clause
60(2)(a) of the Conditions of Contract represents the
value the Engineer attaches to the work items carried out
at the time he is preparing his certificate. The value
of unfixed goods and materials are excluded from the
amount.

> * *The Contract Price Fluctuation Clause used
> in appropriate cases with the I.C.E.
> Conditions of Contract provides for the use
> of the Baxter formula.*

COMMENTARY

Section 7 of the CESMM deals with Method-Related Charges

Some cost elements within an inclusive price related to the in position quantity of finished permanent work may not vary in direct proportion to quantity.   For example, such a price will include a cost element for the labour and materials reflected by the physical entity of the work left on Site.   It will include also the cost element covering the bringing to Site, the setting up and the operation, maintenance and removal of plant, machinery and such like related to the method of executing the work. For similar work the former cost element can be said to vary in direct proportion to quantity, whereas the latter cost element may not.

The foregoing serves to illustrate cases can arise where there is just cause to question the logic and equity of a valuation made by taking the total units of the ad-measured quantities of an item and multiplying them by the inclusive price per unit attached to the particular item in the Bill of Quantities.   With the implied object of providing within the Contract a satisfactory means of adjustment and avoiding contention in such cases, the CESMM allows tenderers the option of sub-dividing prices. They may, subject to the provisions of Section 7 of the CESMM, introduce into the Bill of Quantities priced items covering the elements of cost which relate to methods of executing the work (Method-Related Charges) and price the quantities given in the Bill of Quantities at prices which excludes the element of cost covered by the Method-Related Charges elsewhere given.

Definitions

'Method-Related Charge', means an itemised sum as defined in the Paragraph referenced.   Each Method-Related Charge will be either a :-   7.1(a)

'Time-Related Charge', which denotes a Method-Related Charge which covers work the cost of which is related to the time taken to carry it out.   For example, an itemised sum inserted to cover the work of operating and maintaining a facility or service for a period of time.   7.1(b)

'Fixed Charge', which denotes a Method-Related Charge which is not a Time-Related Charge.   For example, an itemised sum inserted to cover the work of bringing to Site and setting up a facility or service.   7.1(c)

Insertion by a Tenderer

The CESMM, without making it compulsory, permits tenderers to insert for any items they may choose, priced items of Method-Related Charges, insofar as they cover work which relates to methods of execution which are considered not to be proportional to quantity, the cost of which has not been allowed in the prices inserted for the quantified items given in the Bill of Quantities.   7.2

COMMENTARY

Insertion by a Tenderer (cont.

The Bill of Quantities when issued to tender will
include several blank pages for tenderers to use if
they wish to insert Method-Related Charges. The blank
pages are included in the General items (Class A) section
of the document. They are ruled and headed in a manner
similar to those on which the work items are given. The
pages will be included in the document in a position
which ensures that any Method-Related Charges inserted
will follow the order set out for them in Class A of the
Work Classification. See also Paragraph 7.3 of the CESMM
and subsequent Commentary.

Itemisation

Itemisation of Method-Related Charges should follow, where
possible, the order of classification and other requirements as set out in Class A. The item descriptions inserted
should state whether the item covered is a Time-Related
Charge or a Fixed Charge and separate items should be inserted for each. Tenderers, if they so wish, may insert
Method-Related Charges to cover items of work other than
those set out in Class A, subject to Paragraph 7.2 of the
CESMM.   7.3

Descriptions

Paragraph 7.4 of the CESMM establishes the rules that the
description inserted in the Bill of Quantities for a Method-
Related Charge shall define precisely the work covered and
shall identify the resources intended to be used, and shall
identify the particular items of Permanent Works or Temporary
Works, if any, to which the Method-Related Charge relates.   7.4

Contractor Not Bound to Adopt Method

The Contractor is not bound to adopt the intended method
inserted in the description of an item for a Method-Related
Charge.   7.5

Charges Not to be Measured

"Method-Related Charges shall not be subject to admeasurement".   7.6

The quantity of time taken for the work covered by a
Time-Related Charge and the quantity of work and resources
covered by a Fixed Charge is not subject to admeasurement.

Quantities of the time it is anticipated will be taken
for the work covered by a Time-Related Charge are not required
to be stated in the item description. The unit of measurement
for a Method-Related Charge is the sum. The inserted sum being
an amount for the work covered by the Charge and not a rate to
be used to determine the Charge by admeasurement.

An appropriate adjustment to a Method-Related Charge would
be made where instructions, changing the given requirements
upon which the Contractor tendered, made the inserted related
Method-Related Charge unreasonable or inapplicable. In the

COMMENTARY

Charges Not to be Measured (cont.

absence of such instructions the sum inserted for any
Method-Related Charge is not subject to adjustment.

Payments

Method-Related Charges are certified and paid as part of
the Contract price. In his monthly statement (I.C.E
Conditions 60(1)) the Contractor includes the estimated
amounts of the Charges to which he considers himself
entitled and the Engineer includes in his certificate
the amount of the Charges which in his opinion on the
basis of the statement is due to the Contractor.

7.7

The amount due for work covered by a Fixed Charge
is taken as the same proportion of the total Charge as
the work done bears to the total work covered by the
Charge. The total of the work covered by a Fixed Charge
should be apparent from the item and the proportion can
be determined by fact.

From the viewpoint of administering the Contract it
is advantageous to separate Fixed Charges for setting up
services or facilities from those for their removal. The
Instructions to Tenderers should state which Fixed Charges,
if any, are to be inserted in the Bill of Quantities in
this way.

There is nothing in the rules of the CESMM, which
prevents tenderers from including in Fixed Charges for the
establishment of plant and equipment, the capital cost or
depreciation cost intended to be charged in respect of
them to the Contract. Where such costs are included in a
Fixed Charge, the Contractor is entitled to be paid them,
in a certificate at the appropriate time, as soon as the
plant and equipment envisaged by the Fixed Charged is
established. It is often unacceptable that the Employer
should finance the Contractor in this way. Where this is
the case, instructions to tenderers should require that
any allowance tenderers wish to make for the capital cost
of or the depreciation cost for any plant and equipment in a
Method-Related Charge, shall be included in a Time-Related
Charge and not as part of a Fixed Charge.

The amount due for work covered by a Time-Related
Charge is taken as the same proportion of the total Charge
as the relevant period of operation bears to the estimated
total period of operation. Note that the total period of
operation needs to be estimated. If as the work proceeds
the estimate of the total period of operation proves
unrealistic it should be revised. Thus the denominator
of the fraction which determines the proportion may change
from time to time.

COMMENTARY

Payment When Method Not Adopted

Provided the result is satisfactory and provided the
Engineer has not ordered a change of method, the Contractor is ultimately paid the amount attached to the
item inserted for a Method-Related Charge, whether or
not the Contractor uses the method described in the
item.   7.8

Where the method is changed, the timing of the
payment of the Charge, or the unpaid balance of the
Charge where the changed method is substituted for the
inserted method during progress, is a matter for agreement between the Engineer and the Contractor. Failing
such agreement the Charge or the unpaid balance, as the
case may be, is treated as if it were an addition to
the Adjustment Item (See 6.3 and 6.4, Section 6, this
Chapter).

Any change of method ordered by the Engineer would
be dealt with as a variation and would be valued as provided in Clause 52 of the I.C.E. Conditions of Contract.

WORK CLASSIFICATION

The application of the Work Classification set out in
Section 8 of the CESMM, is explained in Section 3 of the
document. (See Previous Commentary on Section 3 in this
Chapter).

Section 8 commences with an index of the twenty four
Classes contained in the Section and subsequently gives a
classification table with Notes appended for each Class.

In subsequent Chapters of this book each Class, or a
group of allied Classes, is made the subject of a separate
Chapter. Within each, Tables are given which summarise in
outline the requirements to ensure compliance with the
Notes of the CESMM when items are formulated from the
descriptive features noted in the particular Table. The
Tables refer only to the requirements imposed by the Notes
in the Work Classification. The full rule for a particular
item must take account of any appropriate requirements in
Sections 1 - 7 of the CESMM in addition to those noted in
the Tables. Commentary which discusses the detail of
applying the requirements of the CESMM to practical measurement and bill compilation is given to complement the Tables.

3 General Items—CESMM Class: A

General items, Class A of the CESMM covers general obligations, site services and facilities, Temporary Works, testing, Provisional Sums and Prime Cost Items. It also provides for the Contractor to include Method-Related Charges, if he so chooses.

Table 3.01 Class A Generally

Generally - The unit of measurement for the General Items shall be the sum, except where another unit of measurement is used in accordance with Note A5 (Note A1)

COMMENTARY

Class A Generally (refer to Table 3.01)

Unlike other Classes of the CESMM, a unit of measurement is not given against the descriptive features in the classification table. Except for certain items of "specified requirements" which may be quantified, the unit of measurement for the Class A items is the sum. (See subsequent Commentary on Specified Requirements).

Table 3.02 Contractual Requirements

1st Division	2nd Division	
Contractual requirements	Give item for performance bond, if required	
	Give separate items for insurance:- of the Works of constructional plant against damage to persons and property The Contractor may insert Method-Related Charges in accordance with Paragraph 7.2 for additional insurances. (Note A2)	The standard features for insurances cover only insurances in accordance with clauses 21 and 23 of the I.C.E. Contract, unless otherwise stated. (Note A2)

COMMENTARY

Contractual Requirements (refer to Table 3.02)

The standard contractual requirements items for insurances given in the CESMM cover insurances required to be effected in accordance with the I.C.E. Contract (Clauses 21 and 23), unless otherwise stated. Any insurances required in addition to, or different in scope from, the insurances prescribed in the I.C.E. Contract, should form the subject of Special Conditions of Contract and additional description or additional items should identify the requirement of the Special Conditions.

Where the Contract provides for the Contractor to indemnify the Employer against stated risks but makes no provision for him to insure against them, the Contractor may if he chooses, insure against the stated risks and include the cost of so doing as Method-Related Charges.

Table 3.03   Specified Requirements

Generally   Items classed as specified requirements are given for work other than the Permanent Works where the Contract expressly states the nature and extent of such work (Note A3)

A quantity shall be given against an item where the value is to be ascertained and determined by admeasurement (Note A5)

Item descriptions for work, other than those for items which are quantified, shall distinguish between:-

establishment and removal of services or facilities, and

continuing operation or maintenance (Note A6)

Item descriptions for plant associated with Temporary Works shall distinguish between:-

plant operating, and

plant standing by (Note A4)

1st Division	2nd Division	3rd Division
Specified requirements	Accommodation for the Engineer's staff	State type of accommodation
	Services for the Engineer's staff	State nature of service
	Equipment for use by the Engineer's staff	State type of equipment
	Attendance upon the Engineer's staff	Identify attendant
	Testing of materials	Give items for all testing not given separately in other Classes (Note A7)
	Testing of the Works	State particulars of samples and methods of testing (Note A7)
	Temporary Works	Describe in accordance with 3rd Division or other appropriate feature

COMMENTARY

Specified Requirements (refer to Table 3.03)

Items classed as specified requirements are given in the circumstances stated in Note A3 but not otherwise. (See "Generally" in the first panel of Table 3.03). Where the Contractor may himself decide the nature and extent of the work other than the Permanent Works, such work does not qualify to be itemised in the Bill of Quantities.

Specified requirements items which it is intended shall be subject to admeasurement are quantified in appropriate units of measurement. Otherwise the unit of measurement attached to Class A items is the sum and they are not subject to admeasurement.

COMMENTARY

Specified Requirements (cont.

   Separate items make the distinction required by Notes A4 and A6. (See
"Generally" in the first panel of Table 3.03. See also the items under the
heading "Specified Requirements" in the specimen of the Part of the Bill of
Quantities given in the Example at the end of this Chapter).

Table 3.04   Method-Related Charges

Method-Related Charges	May be inserted by tenderers in accordance with Section 7 of the CESMM (Note A8)
	Itemisation and descriptions shall be in accordance with CESMM, Paragraphs 7.3 and 7.4, respectively
	Items shall distinguish between Time-Related and Fixed Charges (Note A8)

Table 3.05   Provisional Sums

1st Division	2nd Division	3rd Division	
Provisional sums	Dayworks	Give a separate item each with its associated sum for:-	
		Labour	sum
		Materials	sum
		Plant	sum
		Where a Daywork Schedule as alternative (b) of Paragraph 5.6 of the CESMM is included in the Bill of Quantities, following each of the sums for the foregoing, give an item for percentage adjustment in accordance with the 3rd Division features (Note A9)	%
		The percentage adjustment inserted by tenderer should correspond with that inserted in the Daywork Schedule	
	Other Provisional Sums	Give separate items identifying each of any specified contingencies required with their associated sums	sum

COMMENTARY

Provisional Sums (refer to Table 3.05)

   For "Dayworks" refer to Commentary 5.6 and 5.7, Chapter 2. For "Provisional
Sums" refer to Commentary 5.17, Chapter 2. See also the items in the specimen
of the Part of the Bill of Quantities given in the Example at the end of this
Chapter.

PART 2 : BANDWOOD SITE

Number	Item description	Unit	Quantity	Rate	Amount £	p
	SITE INVESTIGATION					
	Trial holes, Specification clause 12.1					
B110.1	Number; maximum depth 3.5 m.	nr	4			
B140.1	Depth in material other than rock.	m	11			
B150.1	Depth in rock.	m	3			
B170.1	Depth backfilled.	m	14			
	Trial holes, Specification clause 12.2					
B110.2	Number; maximum depth 2.0 m, to explore for pipes, cables or other underground structures.	nr	2			
B140.2	Depth in material other than rock	m	4			
B170.2	Depth backfilled.	m	4			
	Boreholes					
	Diameter 100 mm					
B221	Number in material which includes rock from which cores are to be taken; maximum depth 20 m.	nr	6			
B231	Number in material which includes rock from which cores are not to be taken; maximum depth 20 m, capped.	nr	2			
B241	Depth in material other than rock.	m	40			
B251.1	Depth in rock; open drilling.	m	30			
B251.2	Depth in rock; including recovery of 76 mm diameter cores and logging, packing and transport as Specification clause 15.	m	90			
B261	Depth cased.	m	32			
B271	Depth backfilled.	m	160			
B281	Removal of obstructions.	h	8			

COMMENTARY

Maximum depth of holes are stated in the item descriptions for the number of holes. See Note B2 of the CESMM.

It is thought that the classification 'samples' is inappropriate to continuous rock cores recovered from boreholes. Additional description makes it clear that the work of recovery etc., is included in Item, Code B251.2. Separate items are given to distinguish open drilling from that from which cores are recovered.

(1)   To Part 2 Summary   Page total

PART 2 : BANDWOOD SITE

Number	Item description	Unit	Quantity	Rate	Amount £	p
	SITE INVESTIGATION				COMMENTARY	
	Samples, as Specification clause 16 to 18 (inclusive), logging, packing and transporting as Specification clause 17				Particulars of logging, packing and transport shall be stated, CESMM Note B7	
B420	Undisturbed samples of cohesive material taken from boreholes, diameter 100 mm, length 450 mm.	nr	8		Minimum size of samples shall be stated (in accordance with BS.1377), CESMM Note B7.	
B430	Undisturbed samples of non-cohesive material, taken from boreholes, diameter 100 mm, length 450 mm, retrieved by sampler as Specification clause 18.	nr	2			
B440	Disturbed samples of cohesive or granular material taken from boreholes, mass 25 kg.	nr	6		Method of retrieval shall be stated for undisturbed samples of non-cohesive material, CESMM Note B8.	
B470	Samples of rock taken from cores recovered from boreholes, diameter 76 mm, length 152 mm.	nr	6			
B480	Samples of ground water taken from boreholes, volume 500 ml.	nr	10			
	Site tests				Maximum depth below surface at which tests are made shall be stated, CESMM Note B9.	
B523	Standard penetration to BS. 1377, test No. 19, maximum depth 15 m.	nr	4			
	Laboratory tests					
B627	Uniaxial unconfined compression tests on rock, as Specification clause 18.	nr	6			
B632	Sulphate content, to BS. 1377, test 9.	nr	6			
	Site instrumental observations					
B721	Piezometer, Installations.	nr	2			
B722	Piezometer, Readings.	nr	40			
	(2) To Part 2 Summary			Page total		

5 Geotechnical and Other Specialist Processes—CESMM Class: C

The classification table for Class C provides descriptive features to formulate items for drilling and grouting by injection to alter the properties of soils and rocks, for diaphragm walls, for ground anchors and for sand drains. Grouting associated with tunnels (except that carried out from the ground surface) is excluded from Class C and is included in Class T

Table 5.01 Drilling, Grouting, Materials and Injection

Drilling for Grouting

1st Division		2nd Division	3rd Division
Drilling for grouting:-			For each size give items for:-
through material other than rock or artificial hard material	State diameter of holes (Note C1)	State zone of inclination as 2nd Division features (See Figure C1.)	Number of holes nr Water pressure tests (if required) nr Length of holes stating 3rd Division depth range m Length drilled through previously grouted holes m
through rock or artificial hard material			

Grout Materials and Injection

1st Division	2nd Division		3rd Division
Grout materials and injection	Injections nr	Measure number of injections expressly required at each separate stage of grouting (Note C4)	When in stages state stage length range or actual stage length, as appropriate, in accordance with 3rd Division features
	Packers nr		
	Stated grout materials as 2nd Division features t	State components of mixtures and proportions by mass (Note C2) Mass of grout material not to include mass of water (Note C3)	When not in stages, describe as "Grouting not in stages"

COMMENTARY

Drilling, Grouting, Materials and Injection (refer to Table 5.01)

Measurement divides the work of drilling and grouting into items for the number of holes, the lengths of holes to be drilled, the number of injections of grout material and the mass of the grout to be injected. The items applicable to the work and the details to be given in item descriptions are outlined in Table 5.01.

COMMENTARY

Drilling, Grouting, Materials and Injection (cont.

Number	Item description	Unit	Diagrams Illustrating Angles of Inclination appropriate to Items
	GEOTECHNICAL AND OTHER SPECIALIST PROCESSES		
	Drilling for grouting through material other than rock or artificial hard material		
	Vertical downwards		
C111	Number of holes.	nr	
C115	Length of holes, depth 10 - 20 m.	m	
C116	Length of holes, depth 20 - 40 m.	m	
	At an angle of 0 - 45 degrees to the vertical downwards		
C121	Number of holes.	nr	
C124	Length of holes, depth 5 - 10 m.	m	
C125	Length of holes, depth 10 - 20 m.	m	
	At an angle not exceeding 45 degrees to the vertical upwards		
C131	Number of holes.	nr	
C134	Length of holes, depth 5 - 10 m.	m	
	At an angle up to but not including 45 degrees to the horizontal		
C141	Number of holes.	nr	
C147	Length of holes, depth 42 m.	m	
	Grout materials and injections; grouting not in stages		
C316	Injections.	nr	
C346	Cement/clay grout as Specification clause 3.7.	t	

Note: Angle α in above diagram = up to but not including 45 degrees to the horizontal.

Fig. C1.   Specimen item descriptions and angles of inclination appropriate to standard descriptive features for drilling and grouting.

Table 5.02   Diaphragm Walls

Generally - Use the classification 'diaphragm walls' for walls constructed using bentonite slurry or other support fluid (Note C5)

1st Division	2nd Division		3rd Division
Diaphragm walls	Excavation:- in material other than rock or artificial hard material m3 in rock m3 in artificial hard material m3	State width of walls (Note C6) Define rock in Preamble (Paragraph 5.5) State nature of artificial hard material (Note C6) Separate items are not required for disposal of excavated material (Note C8)	State depth range (up to 30 m deep), or state actual depth when over 30 m deep and when only one depth in one item
	Concrete m3 (Separate items are not required for trimming the faces of the diaphragm walls (Note C8)	State width of walls (Note C6) State mix specification or strength (Note C6) Measure depth excluding any depth of concrete cut off (Note C7) Calculate volume as Notes F10 and F11 (Note C7) Formwork for voids, rebates and fillets in diaphragm walls shall be classed as concrete ancillaries Class G, (Note C11)	
	Bar reinforcement to BS 4449:- mild steel t high yield steel t	Include mass of stiffening, lifting and supporting steel cast in (Note C10)	State diameter as 3rd Division features
	Waterproofed joints sum		
	Guide walls m	Lengths measured shall be those of diaphragm walls (Note C9)	

Table 5.03 Ground Anchors

Generally - Separate items are not required for drilling for ground anchors or for grouting (Note C12)

1st Division	2nd Division		3rd Division

Ground anchors:-
 	
  temporary
 	
  permanent
	For each size give items for:-
	
	  number of anchors, stating whether in clay, in gravel or in rock nr
	
	  length of anchor cables, stating whether in clay, in gravel or in rock m
		Measure lengths of cables between outer faces of anchorages (Note C12)	State working load range or actual working load, as appropriate, in accordance with 3rd Division features and Paragraph 5.14

COMMENTARY

Diaphragm Walls and Ground Anchors (refer to Tables 5.02 and 5.03)

The standard descriptive features and appropriate CESMM Notes for diaphragm walls and ground anchors are noted or referred to in Tables 5.02 and 5.03. The measured Example at the end of this Chapter relates to and provides Commentary on the measurement of diaphragm walls and ground anchors.

Table 5.04   Sand Drains

1st Division	2nd Division		3rd Division

Sand drains
 	
(Separate items are not required for disposal of excavated material (Note C8)
	For each size give items for:-
	
	  number of drains, nr
	
	  depth of over-lying materials, m
	
	  length of drains m
		State depth range or actual depth, as appropriate, in accordance with 2nd Division features and Paragraph 5.14	State diameter range or actual diameter, as appropriate, in accordance with 3rd Division features and Paragraph 5.14

COMMENTARY

Sand Drains (refer to Table 5.04)

The classification "sand drains" covers their complete construction, including
the columns of filter material, the holes to accommodate them and the disposal
of the material resulting from the holes, in accordance with the requirements
prescribed in the contract. The items which require to be given for the drains
are set out in Table 5.04. The length of drains given is the collected length
of the stated diameter range (or diameter) for each depth range (or depth).
Lengths and classification depths are measured from the base of the holes to
the top of the filter material. Items for the depths of overlying material are
measured for any unfilled holes formed between the top of the filter material
and the working surface.

Sand blankets related to sand drains are measured as filling and compaction
to stated depth or thickness (stating the material), as provided in Class E.
Outlet drains associated with sand drains are measured as provided in Classes
I - L. The items for the work in these Classes may be grouped with those for
the sand drains under an appropriate identifying heading in the Bill of Quantities.

Number	Item description	Unit
	GEOTECHNICAL AND OTHER SPECIALIST PROCESSES	
	Sand drains, as Specification clause 52	
	Diameter 400 - 500 mm	
C713	Number of drains.	nr
C723	Depth of overlying material.	m
C743	Length of drains, depth 10 - 15 m.	m

The specimen item descriptions cover only the sand drains.

Actual diameter is stated in place of the range, if only one diameter in one item.

Fig. C2. Specimen item descriptions for sand drains.

EXAMPLE CE.1

Measured Example

An example of taking off the dimensions for diaphragm walls and ground anchors
is given in the measured example which follows.

The dimensions for the general excavations and other work to the basement
are not given in the Example.

50

Original Surface

2000

750

Commencing Surface

Gravel

16000

Reinforced concrete wall

Chase

Ground Anchors

Clay

3000

WALL SECTION 1:200

ABRIDGED BAR SCHEDULE (Each 4m length of wall)		
TYPE AND SIZE	TOTAL No.	LENGTH OF EACH BAR mm
T25	88	10000
T20	88	1200
T20	40	1000
T16	256	10000

100

200

CHASE DETAIL 1:10

2 x 39 No. Anchors

2 x 24 No. Anchors

40000

25000

Basement

2 x 24 No. Anchors

2 x 39 No. Anchors

PLAN 1:500

DIAPHRAGM WALLS. DRG. No. C/D/1

EXAMPLE CE.1

Wall

2/40.00	80.00
2/25.00	50.00
	130.00
4/0.75	3.00
	133.00
Below commencing surface	16.00
	3.00
depth	19.00

Diaphragm walls
Width 0.75 m

Length	133.00	
width	0.75	
depth	19.00	

Excavn. in matl. other than rock or artif hard matl., depth 19m; Comm. Surf. at level of top of diaphram walls (C414

&

Conc. grade 30/20 as Spec. clause 3.12 (C440

Sum	Waterproofed Joints (C470
133.00	Guide walls. (C480

(1)

COMMENTARY

The length used to calculate volume is measured on the centre line of the wall.

The Commencing Surface is stated in the item description for excavation because it is not the Original Surface. (See Paragraph 5.21 of the CESMM).

The width of wall is stated in both the item description for the excavation and that for the concrete. (See Note C6 of the CESMM). The convention of drawing a line across the description column is adopted to indicate the end of the items to which the width classification heading applies.

The length measured for the guide walls is that of the diaphragm walls. (See Note C9 of the CESMM).

DIAPHRAGM WALLS DRG. No C/D/1

Reinforcement.

$$88 \div 4 = 22$$
$$40 \div 4 = 10$$
$$256 \div 4 = 64$$

High y. steel bars to BS. 4449

133/22/	10.00	25 mm diam. X 3.854 Kg/m = _____ Kg. (C467
133/22/ 133/10/	1.20 1.00	20 mm diam. X 2.466 Kg/m = _____ Kg. (C466
133/64/	10.00	16 mm diam. X 1.579 Kg/m = _____ Kg. (C465

Chase

$$\begin{array}{r} 130.00 \\ 4/0.10 = \quad 0.40 \\ \hline 130.40 \end{array}$$

130.40	Fmw. rough fin. for conc. components of constant cross section; three sides of hor. chase 100 x 200 mm in face of diaphragm wall (G184

(2)

COMMENTARY

The quantities and lengths of reinforcement are taken from the abridged bar schedule.

The waste calculations determine the number of bars per metre of wall. This number is used in conjunction with the length of wall in metres as the timesing factors applied to the length of the bars to calculate the total length of each diameter of bar.

The dimensions are set down in a manner which allows total lengths of bars to be multiplied by the mass per metre to calculate the mass for each bar diameter. The mass for each bar classification would be reduced to tonnes for billing.

Formwork for the chase is classified as Concrete Ancillaries and is measured as provided in Class G of the CESMM. (See Note C11 of the CESMM). The formwork in the Example is measured by length as one item in preference to measuring items for each of the separate surfaces. (See Note G4 of the CESMM).

DIAPHRAGM WALLS DRG. NO C/D/1

Ground anchors

<u>Permanent as Spec.
clause 3.17 working
load 20 t.</u>

$$2/39 = 78$$
$$2/24 = \underline{48}$$
$$\underline{126}$$

<u>126</u>		Number of anchors in clay (C612
126/15.00		Length of anchor cables in clay (C622
<u>126</u>		Number of anchors in gravel (C632
126/15.00		Length of anchor cables in gravel (C642

<u>Trial as Spec. clause
3.20, working load
20 t</u>

<u>2</u>   Number of anchors in clay   (C912

(3)

COMMENTARY

Working loads of anchors require to be given in the item descriptions for both the anchors and the length of anchor cables. In the adjoining example it is considered convenient to use a heading incorporating the load appropriate to the items listed under the heading. Actual working load, because there is only one applicable to the items, is given in place of the load range in the Third Division of the classification table in Class C of the CESMM. (See Paragraph 5.14 of the CESMM).

The length of anchor cables is measured between the outer faces of the anchorages.

DIAPHRAGM WALLS. DRG.No.C/D/1

2/	15.00	Trial anchs. (cont.
		Length of anchor cables in clay (C922
	2	Number of anchors in gravel (C932
2/	15.00	Length of anchor cables in gravel (C942

General Items, Specified requirements

Testing of Works

2/	126	Testing perm.g. anchs. (wkg load 20 t) to 1.5 wkg load (A260.1
2/	2	Testing trial anchs. (wkg. load 20t) to failure (A260.2

COMMENTARY

The non-standard descriptive feature "trial anchors" is explicit and is used in the item descriptions in preference to the standard descriptive feature "temporary anchors".

Testing of anchors expressly stated in the Contract to be carried out by the Contractor where the nature and extent of the tests are expressly stated in the Contract, are classified specified requirements and are given as provided in Class A of the CESMM. (See Note A3 of the CESMM).

(4)

6 Demolition and Site Clearance—CESMM Class: D

"Includes" and "Excludes" preceding the classification table in the CESMM make clear the intention that Class D applies to those parts of subjects to be demolished or cleared which are above the Original Surface. Work below the Original Surface will normally be of a kind which it is appropriate to measure in accordance with Class E. Strict application of these conventions is not always convenient. The Specification should make clear the level down to which buildings and structures are to be demolished. It may define this as the top of the lowest slab or another datum which may be below the Original Surface. Where this is the case, item descriptions for demolition should state the work below the Original Surface which is intended to be included in the demolition items. See Note D7 of the CESMM.

The volume of a building or structure measured is that of the solid the part to be demolished would create above the Original Surface if it had no internal voids. This is the volume intended by the CESMM to be stated in the item descriptions for the demolition of buildings and structures, whether or not they include parts which are below the Original Surface. See Note D6 of the CESMM.

Where demolition abuts construction which is to remain, consideration must be given to the method of dealing with any support, protection, reinstatement and making good required to the remaining construction. Simple detachment one from the other may be covered by the Specification and a note in the Preamble of the Bill stating that prices for demolition are to include for such work and any incidental making good. More involved work of this kind, where it is considered a reasonable estimate of requirements can be made, may be given as measured items appropriately classified. Where it would be difficult prior to demolition to assess the nature and extent of the work to the existing it is appropriate to make it the subject of a Provisional Sum.

Where shoring is specified to support the remaining construction it is appropriate to consider that of a temporary nature as Temporary Works. That which is to be left in position on completion of the Contract being considered as part of the Permanent Works. Where the Contractor may himself select the means of complying with the requirements for temporary shoring, no items are given for it in the Bill of Quantities. Where temporary shoring is precisely specified and detailed it is given as items of specified requirements under Class A. The Engineer will usually design the shoring where it is to be left in position on completion of the Contract. Such shoring may be measured and given under an identifying heading in the Bill or it may be given as a Provisional Sum.

Where specific requirements are imposed in regard to hoardings, overhead platforms, fans or other temporary works related to the demolition, they need to be given as items of specified requirements under Class A.

Where work of diversion, removal, disconnection or sealing off of Statutory Authority's services is required in connection with demolition, this can only be undertaken by or under the order of the appropriate Authority. The Specification will usually require the Contractor to give the necessary notices and arrange for the work to be carried out. This is covered in the Bill of Quantities by giving a Provisional Sum for the work itself and for any related fees which may be repayable under Clause 26(1) of the I.C.E. Contract.

Table 6.01 indicates descriptive features and units of measurement and refers to the Notes in the CESMM. Commentary on Class D is included in the Measured Example at the end of this Chapter.

Table 6.01   Demolition and Site Clearance

Generally - State when materials arising are to remain property of Employer (Note D3)

1st Division		2nd Division	3rd Division
General clearance ha (See Commentary to column (1) of dimensions in subsequent Measured Example this Chapter)	Includes demolition and removal of all articles objects and obstructions which are expressly required to be cleared except those for which separate items are given (Note D1) Identify area to be cleared if it is not the total area (Note D2)	State type of land:- Urban land Agricultural land Woodland Moorland	
Trees and stumps nr Measure only trees exceeding 500 mm girth and stumps exceeding 150 mm diameter	Measure girth of trees one metre above ground (Note D4) Separate items are not required for stumps of trees which are themselves removed (Note D4)	State girth range as 2nd Division features. State actual girth if only one in one item	
Buildings sum Other Structures sum	Identify buildings and other structures (Note D5) An identified group of buildings or other structures may be given as a single item (Note D5) Parts of buildings and structures below the Original Surface are excluded unless otherwise stated (Note D7)	State predominant material:- Brickwork Concrete Masonry Timber No predominant material	State volume range as 3rd Division. State actual volume if over 5000 m3 or if only one in one item. The volume stated shall be the approximate volume occupied excluding any volume below the Original Surface (Note D6)
Pipelines m Separate items not required for demolition of supports (Note D8)	Measure pipelines when above ground and over 100 mm nominal bore Measure pipelines within buildings and structures only when nominal bore exceeds 300 mm (Note D8)	State nominal bore range as 2nd Division features. State actual nominal bore if only one in one item	

EXAMPLES DE.1 and DE.2

Measured Example (DE.1) - Specimen Bill (DE.2)

The Measured Example which follows is for the site clearance and demolition shown in Drawing No. D/D/1. An example of the Bill of Quantities for the work is given as Example DE.2.

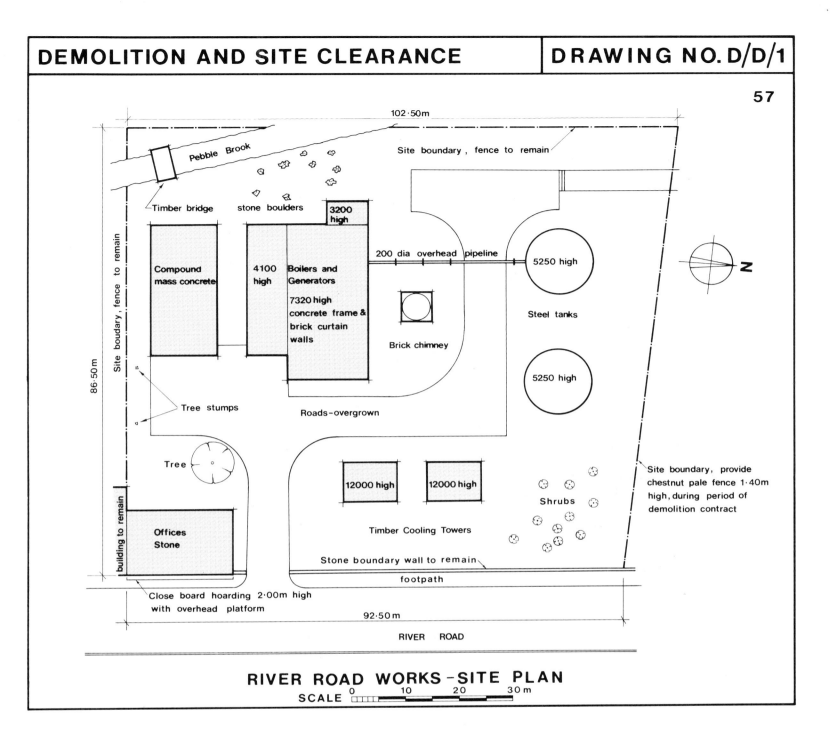

RIVER ROAD WORKS - SITE PLAN

SCALE | 0 | 10 | 20 | 30 m

Pebble Brook

Site boundary, fence to remain

102·50m

Timber bridge

stone boulders

3200 high

200 dia overhead pipeline

5250 high

Steel tanks

Compound mass concrete

4100 high

Boilers and Generators

7320 high concrete frame & brick curtain walls

Brick chimney

5250 high

Site boudary, fence to remain

86·50 m

Tree stumps

Roads-overgrown

Tree

12000 high

12000 high

Shrubs

Site boundary, provide chestnut pale fence 1·40m high, during period of demolition contract

building to remain

Offices Stone

Timber Cooling Towers

Stone boundary wall to remain

footpath

Close board hoarding 2·00m high with overhead platform

92·50m

RIVER ROAD

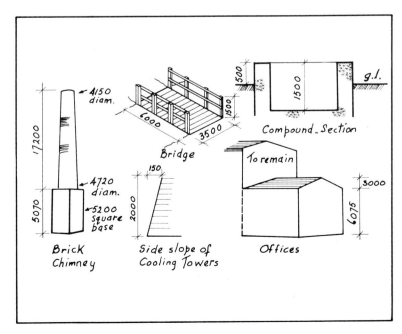

4150 diam.

17200

4720 diam.

5070

5200 square base

Brick Chimney

6000

3500

Bridge

1500

1500

500

1500

g.l.

Compound Section

150

2000

Side slope of Cooling Towers

To remain

3000

6075

Offices

PLAN DIMENSIONS OF BUILDINGS & STRUCTURES		
Ref.	m	
Compound	25.00 x 12.50	See sketch.
Boils + Gen :-		
main buildg.	30.00 x 15.00	
South annexe	25.00 x 7.50	
West annexe	7.50 x 5.00	
Chimney	5.20 x 5.20	See sketch
Steel tanks	12.50 diam	both identical
Cooling towers	10.00 x 7.50	See sketch both identical
Offices	20.00 x 12.50	See sketch
Bridge (stream)	6.00 x 3.50	See sketch

OTHER PARTICULARS		
O/h. pipeline — length = 29.50 m		
Tree (W. offices) = 1200 girth (1m above g.l.)		
Stumps (" ") = both 300 mm diam.		

SITE SKETCHES AND NOTES

EXAMPLE DE.1

<u>Site clearance</u>

102.50
92.50
2)195.00
97.50

97.50		
86.50	8433.75	Gen. clearance, urban land (Site area excluding areas of buildgs & strucs to be demolished. (D110

25.00		Ddt. ditto.
12.50	312.50	(compound.
30.00		
15.00	450.00	
25.00		boils. and
7.50	187.50	generators
7.50		
5.00	37.50	
5.20		
5.20	27.04	(chimney
2/22/7/ 6.25		2)12.50
6.25	245.54	6.25 (steel tanks
2/ 10.00		
7.50	150.00	(coolg. towers
20.00		
12.50	250.00	(offices
37.50		
5.00	187.50	(stream
	1847.58	
	6586.17	Demolition, etc.

$$\frac{2 \times 12.00 \times 0.15}{2.00} = 1.80$$

10.00 × 7.50
1.80
8.20 5.70
10.00 7.50

coolg. towers   2)18.20 × 13.20
mean dimensions   9.10 × 6.60

(1)

COMMENTARY

On the premise that tenderers need to obtain the information they require to prepare their tender from an inspection of that which is to be demolished or cleared, the CESMM does not call for elaborate descriptions of the work. The descriptions are, however, required to identify the buildings and structures to be demolished. A site plan similar to Drawing No. D/D/1 will usually accompany the Bill of Quantities.

The item of general clearance is quantified in terms of the area of the site to be cleared and is given in the Bill of Quantities in hectares.

General clearance includes the removal of trees of a girth not exceeding 500 mm (trunk girth measured 1 m above ground level), the removal of stumps of trees not exceeding 150 mm diameter, the removal of pipelines which are above the ground and which do not exceed 100 mm nominal bore and the removal of anything else which does not qualify to be separately itemised under Class D, but is expressly required to be removed.

When it is not the whole Site which is to be cleared, the item description for general clearance identifies the area to be cleared. See Note D2 of the CESMM.

The waste at the foot of column (1) of the dimensions is the commencement of the preliminary calculations for use when calculating the volumes for subsequent items of demolition.

Demolition, etc. (cont.

```
                        4150
                        4720
        chimney    2) 8870
        mean diam = 4435
        mean radius 2218
```

SCHEDULE OF VOLUMES

	Length	Width	Height	Vol (m3)
Compound.	25.00 ×	12.50 ×	0.50 =	156.25
Boils	30.00 ×	15.00 ×	7.32 =	3294.00
of	25.00 ×	7.50 ×	4.10 =	768.75
Generators	7.50 ×	5.00 ×	3.20 =	120.00
				4182.75
Chimney $\frac{22}{7}$/	2.22 ×	2.22 ×	17.20 =	266.42
	5.20 ×	5.20 ×	5.07 =	137.09
				403.51
Tanks 2/$\frac{22}{7}$/	6.25 ×	6.25 ×	5.25 =	1289.06
C/towers 2/	9.10 ×	6.60 ×	12.00 =	1441.44
Offices	20.00 ×	12.50 ×	6.08 =	1520.00
$\frac{1}{2}$/	20.00 ×	12.50 ×	3.00 =	375.00
				1895.00
Bridge	6.00 ×	3.50 ×	1.50 =	31.50

Sum   Strucs. concrete, vol. 156 m3; Compound adjoing. S. boundary, including demolishing walls to 1 m below Orig Surf. (D423

Sum   Buildgs. concrete fra of b.kw. walls, vol. 4183 m3; Boils of Gen. (D397

(2)

COMMENTARY

The items for demolition assume the Specification calls for demolition down to the top of the lowest solid ground slab. It is assumed also, with the exception of the Compound, that the top of the lowest slab is at or above the Original Surface. The heights used to calculate the "volume occupied" are measured, therefore, from the top of the lowest ground slab.

The item description for the Compound indicates there is demolition below the Original Surface. See Note D7 of CESMM.

Buildings and structures are identified in the item descriptions. See Note D5 of the CESMM.

Each item is for a specific building, structure or group. Consequently there will be but one volume in each item and the actual volume is stated in the item description in place of the range. See Paragraph 5.14 of the CESMM. See also previous Commentary on this Paragraph in Chapter 2.

It is a matter of personal preference whether a schedule of volumes is prepared or whether the dimensions for each building or structure are set down preceding each item.

There is no standard classification for the composite construction of the Boilers and Generators building. It is appropriate to use the digit 9 in the second division position in the Code to denote this. See Paragraph 4.5 of the CESMM.

DEMOLITION AND SITE CLEARANCE   DRG. NO. D/D/1

Demolition etc. (cont.

<u>Sum</u>	Strucs. bkw., vol 404 m3; Chimney. (D414	
<u>Sum</u>	Group of two strucs., metal, vol 1290 m3; two Steel Tanks (D446	
<u>Sum</u>	Group of two strucs., timber, vol. 1441 m3; two Cooling Towers. (D456	
<u>Sum</u>	Buildgs. masonry, vol. 1895 m3; Offices. (D336	
<u>1</u>	Trees of girth 1.20 m. (D220	
<u>2</u>	Tree stumps of diam. 300 mm. (D260	
<u>29.50</u>	Pipelines, nom. bore 200 mm. (D510	

Work affected by bodies of water

<u>Sum</u>	Strucs. timber, vol. 32 m3; Bridge over Pebble Brook. (D451	

COMMENTARY

An identified group of buildings or structures may be given as a single item. See Note D5 of the CESMM.

The unit of measurement for buildings and structures is the sum.

Trees exceeding 500 mm girth (trunk girth measured 1 m above ground level), are given by number, stating the girth or the girth range.

Stumps of trees exceeding 150 mm diameter are given by number stating the diameter or the diameter range.

Pipelines which exceed 100 mm nominal bore which are above ground are measured linearly in metres stating the nominal bore or the nominal bore range. Separate items are not required for the removal of supports.

It is assumed in the Example that there are no pipelines exceeding 300 mm nominal bore within the buildings and structures to be demolished.

The actual girth of the tree, the actual diameter of the stumps and the actual nominal bore of the pipeline each represent a single dimension in one item and are given in the item descriptions in place of the range. See Paragraph 5.14 of the CESMM.

Work affected by bodies of water is distinguished as provided in Paragraph 5.20 of the CESMM.

The Example includes only Class D items. Specified requirement items and items for the work to the remaining buildings have not been included.

(3)

PART 3 : CLEARANCE OF SITE

Number	Item description	Unit	Quantity	Rate	Amount	
					£	p
	DEMOLITION AND SITE CLEARANCE					
	General clearance					
D110	Urban land; Site area excluding areas of buildings and structures to be demolished.	ha	0.66			
	Trees and stumps					
D220	Trees of girth 1.20 m.	nr	1			
D260	Tree stumps of diameter 300 mm.	nr	2			
	Buildings					
D336	Masonry, volume 1895 m3; Offices.	sum				
D397	Concrete frame and brickwork walls, volume 4183 m3; Boilers and Generators.	sum				
	Structures					
D414	Brickwork, volume 404 m3; Chimney.	sum				
D423	Concrete, volume 156 m3; Compound adjoining South boundary, including demolishing walls down to 1 m below Original Surface.	sum				
D446	Group of two, metal, volume 1290 m3; two Steel Tanks.	sum				
D456	Group of two, timber, volume 1441 m3; two Cooling Towers.	sum				
	Pipelines					
D510	Nominal bore 200 mm.	m	30			
	Work affected by bodies of water					
D451	Structures, timber, volume 32 m3; Bridge over Pebble Brook.	sum				
	(1) To Part 3 Summary Page total					

7 Earthworks—CESMM Class: E

Earthworks covered by Class E includes excavation, dredging, filling, compaction and landscaping. Excavation which is excluded from the Class is listed in the "Excludes" at the head of the Class E classification table in the CESMM

Table 7.01 Excavation

Generally – Excavation is measured nett – No allowance is made for working space (Note E8) – Disposal is included in the excavation items (Note E9) – Separate items are not required for (i) Upholding sides (ii) Keeping free from water (Note E9).

Give item for each separate stage of excavation where separate stages in conduct of work expressly required (Note E7).

Identify Commencing Surface if it is not Original Surface. Identify Excavated Surface if it is not Final Surface (Section 5, Paragraph 5.21).

An isolated volume of rock or artifical hard material in excavation is not measured separately unless its volume exceeds one cubic metre (Note E13)

1st Division

Dredging	m3	State if measured from other than soundings. (Note E11). State location and limits if not clear (Note E1).
Excavation of cuttings	m3	Includes excavation below embankments (Note E2)

Excavation of foundations	m3	State location and limits if not clear. State if around pile shafts (Note E3)	See Note E8 for the volume measured for structures and foundations.
General excavation	m3	Includes borrow pits expressly required (Note E4)	

2nd Division

– Material to be excavated – Deemed to be natural material other than top soil, rock or artificial hard material unless otherwise stated (Note E6)

Top soil Material (See above) Rock (define in Preamble Section 5, Paragraph 5.5)	for re-use	All material deemed to be for re-use unless stated for disposal (Note E5)
Artificial hard material exposed at Commencing Surface (State nature of material – Note E6).	for disposal on site	State location of disposal (Note E5)
Artificial hard material not exposed at Commencing Surface (State nature of material – Note E6)	for disposal	Means disposal off Site unless otherwise stated (Note E5)

3rd Division – Applicable only to Excavation of foundations and General excavation

Maximum depth:	State range of depth in which the maximum depth of the particular excavation occurs.(Measure from Commencing Surface to Excavated Surface – Section 5 Paragraph 5.21)

Table 7.02   Excavation Ancillaries

1st Division - Excavation ancillaries

2nd Division			3rd Division
Trimming of slopes	m2	Measure only for sides of excavation inclined at an angle over 10 degrees to the horizontal and required to be trimmed, not where left as excavated. (Note E26)	Describe material in accordance with 3rd Division. State the precise nature of the material for features 3 and 4
Preparation of surfaces	m2	Measure only where Permanent Works, other than earthworks, are in direct contact with excavation. (Note E27).	
Double handling of excavated material	m3	Measure only if expressly required. Volume is that of void in stockpile. (Note E14).	
Excavation of material below Final Surface and replace with stated material	m3	State the precise nature of the replacement material.	
Timber supports left in	m2	Measure area of supported surface for which the supports are expressly required to be left in (Note E15).	
Metal supports left in	m2		
Dredging to remove silt	m3	Measure only when expressly required that silt which accumulates during period of maintenance shall be removed. (Note E12).	

COMMENTARY
___

Excavation (refer to Table 7.01)
___

Excavation is measured by volume and the quantities are given in the Bill of Quantities in m3. Descriptive features for excavation and a summary of the CESMM Notes are given in Table 7.01.

Excavation is measured the nett size of the excavated void with no allowance for working space. The volume being that before the material is excavated. The convention adopted for the measurement of excavation to accommodate structures is illustrated later.

Excavation items do not include filling and compaction. Where the operations include filling and compaction with material excavated on the site, there must be two items, one for excavation and one for filling and compaction. In some cases there will be a third item of double handling where double handling of the excavated material is expressly required. See "Double Handling" later in the Commentary.

COMMENTARY

Excavation (cont.

The item descriptions for excavation state what is to happen to the excavated material when it is excavated and a descriptive feature in the items will describe whether the material is "for re-use", or "for disposal" (meaning disposal off site), or "for disposal on site". Where excavation items describe the excavated material as "for disposal on site", the location of the disposal must be stated in the item description. If, in relation to material for disposal on site there is a spreading or other deposition requirement, it must be identified in the item description. If there is a definite filling requirement, such as spreading of the material to designed profiles, even with minimum compaction, a filling and compaction item in addition to the excavation item should be measured.

Unless otherwise stated in the item descriptions, all excavation is deemed to be in natural material, other than top soil or rock, and all excavated material is deemed to be for re-use.

It is not necessary to state a Commencing Surface or an Excavated Surface when they are the Original Surface and the Final Surface, respectively. The excavation in an item is, therefore, deemed to start at the Original Surface and finish at the Final Surface when a Commencing Surface or an Excavated Surface is not identified in the item description.

The item descriptions "Excavation of cuttings" or "Excavation of foundations", or "General excavation", without further wording, each denote by application of the provisions in the two preceding paragraphs, that the material to be excavated is natural material other than top soil or rock, that it is for re-use and that the excavation in the item starts at the Original Surface and finishes at the Final Surface.

Items will separate excavation according to the material to be excavated. Horizontal bands of different material are not normally measured as stages of excavation, unless their excavation in stages is expressly required. Exceptionally, consideration may be given to measuring the excavation in stages where it is thought prices would be influenced by the depth in the excavation of a band of particular material.

There is no requirement for excavation in running sand to be separately itemised and it will be classified as in natural material other than top soil or rock, unless the discretion afforded by paragraph 5.10 of the CESMM is exercised and it is decided to make the excavation in running sand the subject of separate items.

Isolated volumes of rock or artificial hard material which are each less than one cubic metre in volume are not measured separately from the material in which they occur.

Work for which separate items are not required are noted in the first panel of Table 7.01.

COMMENTARY

Excavation (cont.

Excavation of Cuttings

Excavation of cuttings is measured by volume and is given in m3 in the Bill
of Quantities. The depth of excavation of cuttings is not required to be
stated in the item descriptions. The implications of the unqualified item
description, "Excavation of cuttings", is explained in the Commentary on
Excavation Generally.

  Specimen item descriptions for the excavation of the cutting shown in
cross-section in the diagram, Figure E1, are given at the side of the diagram.
The specimen item descriptions assume (i) the excavation starts at the
Original Surface and finishes at the Final Surface (ii) there is no top soil
at the Original Surface (iii) all material excavated is for disposal. The
quantity which would be attached to an item would be the volume of the particular class of material to be excavated.

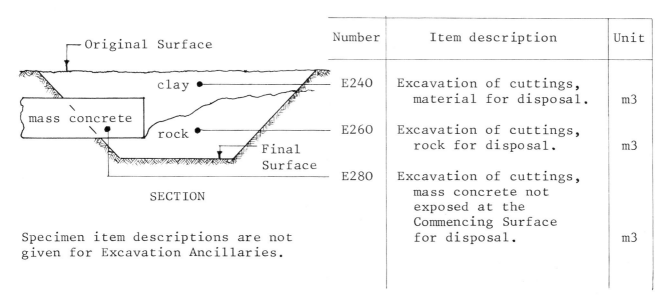

Number	Item description	Unit
E240	Excavation of cuttings, material for disposal.	m3
E260	Excavation of cuttings, rock for disposal.	m3
E280	Excavation of cuttings, mass concrete not exposed at the Commencing Surface for disposal.	m3

Specimen item descriptions are not
given for Excavation Ancillaries.

Fig. E1. Excavation of cuttings.

The excavation of cuttings will sometimes require to be measured in stages.
It is common experience for a specification to require that the bottom
150-300 mm of the excavation is left, to protect the formation, and excavated
immediately prior to the laying of the base material. If this is an express
requirement (rather than a condition of use of the bottom surface of the
cutting by the contractor's construction traffic), it constitutes a stage of
excavation for which a separate item must be given. Similarly, where there is
top soil and it is specified to be excavated separately from other material,
the excavation of the top soil is a stage of excavation for which a separate
item must be given.

  The item descriptions for each stage of excavation must identify the
Commencing Surface or the Excavated Surface where these are not also the
Original Surface or the Final Surface, respectively.

COMMENTARY

Excavation (cont.

Excavation of Cuttings (cont.

    Specimen descriptions for the excavation of the cutting, expressly required to be excavated in stages, shown in cross-section in the diagram, Figure E2 are given at the side of the diagram. The specimen item descriptions assume, (i) top soil 150 mm deep (ii) remaining material is natural material, other than top soil or rock (iii) the last 150 mm of excavation is left to protect the formation and is to be excavated as a separate stage.

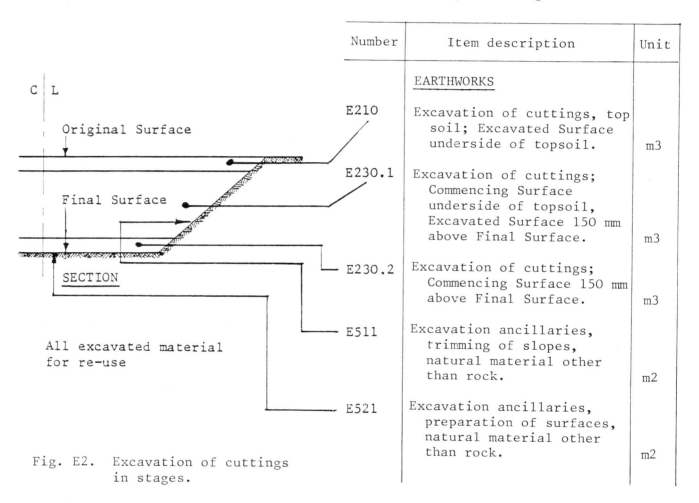

Number	Item description	Unit
	EARTHWORKS	
E210	Excavation of cuttings, top soil; Excavated Surface underside of topsoil.	m3
E230.1	Excavation of cuttings; Commencing Surface underside of topsoil, Excavated Surface 150 mm above Final Surface.	m3
E230.2	Excavation of cuttings; Commencing Surface 150 mm above Final Surface.	m3
E511	Excavation ancillaries, trimming of slopes, natural material other than rock.	m2
E521	Excavation ancillaries, preparation of surfaces, natural material other than rock.	m2

Fig. E2. Excavation of cuttings in stages.

Excavation for Walls at Sides of Cuttings

When walls are required at the sides of cuttings, there is a question of what dividing line to take between the excavation of the cutting and the excavation to accommodate the wall and footings.

    It is suggested that the excavation to accommodate the wall and its backfilling is classified as general excavation, the depth of the excavation being taken down to the formation of the pavement abutting the face of the wall. The excavation below the pavement formation to accommodate the footings being classified as excavation of foundations. The suggested convention is illustrated in cross-section in the diagram, Figure E3.

COMMENTARY

Excavation (cont.

Excavation for Walls at Sides of Cuttings (cont.

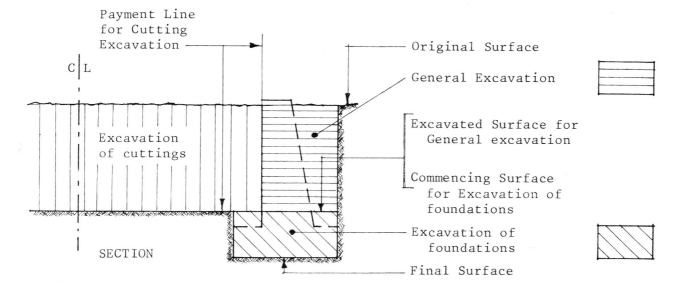

Fig. E3. Excavation for walls at sides of cuttings.

Whatever convention is adopted for dividing and classifying the excavation in this situation the limits of the excavation in any one item will not be clear in the Bill without some additional detail. It is suggested this additional detail could be provided by a diagram on the drawings indicating the measurement convention used for measuring the excavation when preparing the Bill of Quantities. The CESMM requires that item descriptions for "Excavation of foundations" must state the location and limits of the excavation where this is not clear.

Excavation Below Embankments

Excavation below embankments is classed "Excavation of cuttings".

Where an embankment is to be constructed on steeply sloping ground, the surface to receive the embankment may be specified to be benched in steps or trenched. It is suggested that, whilst the excavation for the steps or trenches can be classified as "Excavation of cuttings", it should be identified either by reference to the particular location or by adding to the standard description, the phrase "benched in steps below embankments" or "trenched below embankments", as the case may be.

General Excavation and Excavation of Foundations

The classification "General excavation" covers excavation to reduce levels over areas and will also be used for excavation to reduce a site to the formation level of a structure. "Excavation of foundations" is that which accommodates foundations.

The item descriptions for both these types of excavation must state the range of depth in which the maximum depth of the particular excavation work occurs. Both types are measured by volume and the unit of measurement is m3.

COMMENTARY

Excavation (cont.

General Excavation and Excavation of Foundations (cont.

   Specimen item descriptions for the excavation of foundations and to reduce the levels for the structure shown in cross-section in the diagram, Figure E4, are given at the side of the diagram.

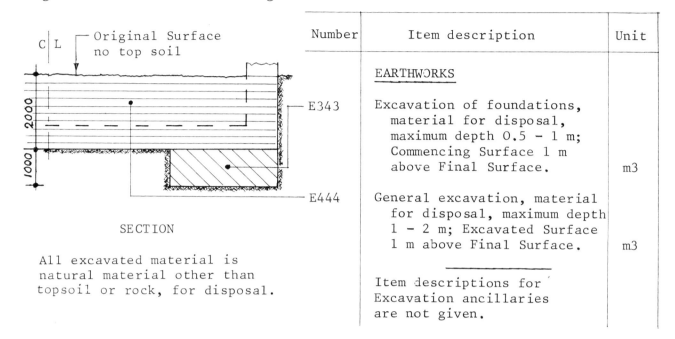

Number	Item description	Unit
	EARTHWORKS	
E343	Excavation of foundations, material for disposal, maximum depth 0.5 – 1 m; Commencing Surface 1 m above Final Surface.	m3
E444	General excavation, material for disposal, maximum depth 1 – 2 m; Excavated Surface 1 m above Final Surface.	m3
	Item descriptions for Excavation ancillaries are not given.	

All excavated material is natural material other than topsoil or rock, for disposal.

Fig. E4. General excavation and excavation of foundations.

Excavation for Structures and Foundations

The volume measured for the excavation of a structure or foundation is the volume either occupied by or vertically above any part of the structure or foundation. Additional working space is not measured. (Note E8 of the CESMM).

   The application of this Note to three situations is illustrated in the following cross-section diagrams. The boundaries of the excavation are shown by dotted lines.

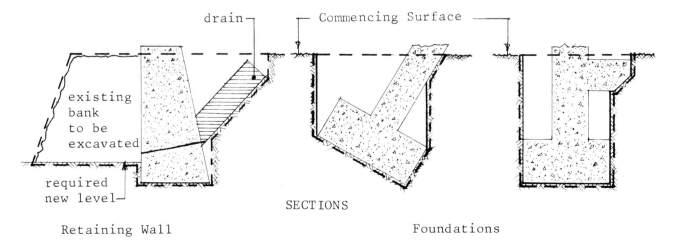

Fig. E5. Volume of excavation measured for structures and foundations.

COMMENTARY

Excavation (cont.

Dredging

The quantities of dredging measured from soundings are the nett in situ volume of the voids formed, given in m3 and calculated from the areas of dredging and the depths obtained from soundings taken before and after dredging. Where quantities are prepared in accordance with this method an hydrographic survey prepared from soundings will usually be provided and the dredging requirements will be plotted on the survey. Reference to this document in the item descriptions or via a clause in the specification, will indicate the location and limits of the dredging.

An alternative method of measuring dredging is to measure the volume of dredged material from the hopper or barge in which the material is initially deposited.

When dredging is measured from other than soundings, the method must be stated in the item descriptions. Excavation classed as dredging in the Bill of Quantities is admeasured as dredging irrespective of the method of excavation adopted by the Contractor (See Note E1 of the CESMM).

Excavation Ancillaries (refer to Table 7.02)

Trimming of Slopes

Where surfaces of earthworks are shown on the drawings or otherwise specified to be sloped to a required angle of over 10 degrees to the horizontal, the surface areas of the slopes, given in square metres, are classified as "Trimming of slopes", and the material to be trimmed is given in the item description.

The surfaces of excavation sloped back during excavation operations and where no work is required to the surface to form them to specified slopes, are not measured.

Preparation of Surfaces

Preparation of surfaces is measured superficially in square metres to the extent that earthworks are to receive Permanent Works other than earthworks. The item covers merely preparing the surface to the required profiles and configurations. It would not include any material or additional operations, such as the application of weedkiller or a surface dressing. These would be given as separate items or, if preferred, may be given in the preparation item by added description.

It is considered consistent with Paragraph 5.10 of the CESMM, to give the preparation of horizontal, vertical and sloping surfaces each as separate items and to add the appropriate word to the standard description.

The nature of the particular material of the surface to be prepared is given in the item descriptions, in accordance with descriptive features.

COMMENTARY

Excavation Ancillaries (cont.

Double Handling of Excavated Material

A separate item for double handling excavated material is measured only when double handling is expressly required. Consequently, excavation and filling items are deemed to include all double handling that may be necessary to carry out the earthworks. The inclusion in the Bill of Quantities of a separate item for double handling excavated material is not decided on whether it would be impossible not to, or whether it is sensible, convenient or practical to stockpile and double handle excavated material, these are matters which are left for the Contractor to decide and allow for accordingly. The measurement of double handling of excavated material is decided by the dictates of the specification or the order of the Engineer.

Double handling of excavated material is measured by volume and is given in the Bill of Quantities in m3. Note E14 of the CESMM requires that the volume shall be that of the void formed in the stockpile. Consequently, when calculating the volume of double handling of excavated material for the quantities in the Bill of Quantities, if the calculations start with the in place volume of the filling for which the material is to be used, an allowance will need to be made for the difference in bulk of the material in the stockpile and its bulk when used as filling.

When calculating the volume of material for double handling on the site the volume of the stockpile will need to be measured both before and after the material has been removed.

Excavation of Material Below Final Surface and Replacement with Stated Material

Soft spots below the formation are among the items to which the classification in the above heading will apply. The precise nature of the replacement material must be given in the item description. The unit of measurement is cubic metres.

Supports Left In

Timber supports and metal supports which are expressly required to be left in are measurable items and are each given separately. The areas given in the items are those of the surfaces supported given in square metres. The dictates of the specification or the order of the Engineer decides when the items are to be measured and included in the Bill of Quantities.

Dredging to Remove Silt

Dredging to remove silt is applicable only to silt which accumulates during the maintenance period and is expressly required to be removed.

The quantities for the Bill of Quantities will require to be estimated because the amount which will accumulate will not be known until measured on completion.

Table 7.03 Filling and Compaction

Generally – Separate items are not required for compaction (Note E18)			
1st Division			
Filling and compaction		The volume of filling and compaction shall be that of the compacted filled volume (Note E19)	
		The volume of excavated material used for filling is deemed to form the same volume of compacted fill (Note E19)	
		State compaction requirements when different compaction requirements are specified for the same material (Note E18)	
		State limitations when rate of deposition is limited (Note E22)	
2nd Division			
To structures	m3	Measure only to the extent that the volume filled is also measured as excavation in accordance with Note 8 (Note E17)	The volume of imported filling is the difference between the total volume of filling material and the volume of excavated material used for filling (Note E19)
Embankments	m3		
To stated depth or thickness	m2	State actual compacted thickness (Note E24)	
		State if over 10 degrees to horizontal (Note E28)	
General	m3		
Pitching	m2	State nature and dimensions of materials (Note E25)	Bulk fill in layers is not classed as to stated thickness (Note E24)
		State actual compacted thickness (Note E25)	
		State if over 10 degrees to horizontal (Note E28)	
Additional filling and compaction necessitated by settlement or penetration into underlying material is measured only to the extent that its depth exceeds 75 mm (Note E19)			
3rd Division – Filling material – Deemed to be unselected excavated material unless otherwise stated (Note E16)			
Excavated topsoil			Where in water and quantities cannot be measured satisfactorily, measure in transport vehicles (Note E21)
Imported topsoil			
Non-selected excavated material other than topsoil or rock			
Selected excavated material other than topsoil or rock			
Imported natural material other than topsoil or rock			
Excavated rock		Filling material shall be classed as rock and excavated rock shall be classed as for re-use only where the use of rock as filling at stated locations is expressly required. (Note E23).	Where in soft areas measure volume in transport vehicles (Note E20)
Imported rock			Where in water and quantities cannot be measured satisfactorily, measure in transport vehicles (Note E21)
Imported artificial material			

Table 7.04   Filling Ancillaries

1st Division – Filling ancillaries

2nd Division			3rd Division
Trimming of slopes	m2	Measure only for sides of filling inclined at an angle over 10 degrees to the horizontal. (Note E26).	Describe material in accordance with 3rd Division features.
Preparation of surfaces	m2	Measure only to extent that surfaces are to receive Permanent Works other than earthworks. (Note E27).	State the precise nature of the material for features 3 and 4.

COMMENTARY

Filling and Compaction (refer to Table 7.03)

Filling is measured as equal in volume to that of the solid which represents the size and shape of the filled volume. The filled volume is taken as equal to the excavated volume (measured before excavation), where the material excavated is used for filling. Additional filling and compaction needed because of settlement of or penetration into the surface upon which the filling is placed is measured where the depth of settlement or penetration exceeds 75 mm.

Item descriptions identify compaction requirements, they state the compaction requirement where different compaction requirements are specified for the same material. The classification "to stated thickness" is not applicable to bulk filling compacted in layers.   See Note E24 of the CESMM. Any limitation on the rate of deposition of filling is stated in item descriptions.

Filling material is deemed to be unselected excavated material other than topsoil or rock unless otherwise stated in the item descriptions. Rock as filling material and excavated rock for re-use as filling are measured only when their use as filling at stated locations is expressly required.

When filling and compaction with material excavated on site is measured, it is necessary to ensure that the filling requirement is matched in volume by that of the excavation classified as in material for re-use. If there is an insufficient volume of excavation to meet the filling requirement, the deficit will, obviously, need to be obtained from another source.

During the measurement process it is not always possible to gauge what the outcome will be when the volume of filling with excavated material is balanced against that of the excavation. A procedure must be instituted or a measurement convention adopted to indicate the position before the excavation and filling items are billed. The procedure used in Example EE.1 classifies all material as for disposal when measuring the excavation initially. A volume of excavation equal to that of any filling with excavated material is re-classified for re-use when filling is measured. If this convention is followed any deficit of excavated material for filling is disclosed by a negative quantity arising when excavation for disposal is cast to a total for billing. Where this happens, excavation and filling items need to be adjusted and correctly classified. An alternative procedure is to leave excavation items to be classified as either for disposal or for re-use until they can be correctly classified by comparison with the filling requirement.

COMMENTARY

Filling and Compaction (cont.

Filling and Compaction to Structures

The classification "filling and compaction to structures" covers filling and compaction to voids excavated for structures and foundations and also filling and compaction within, around and over structures. The diagram in Figure E6 illustrates situations where, if required to be filled as shown, the filling and compaction would be classified as "to structures". The banks marked (a) may suggest the classification "embankments" but banking up of this description is considered to be correctly classified as to structures.

The volume of filling and compaction in an excavated void is restricted to the volume of excavation, measured in accordance with Note E8, which the filling replaces. For example. if reference is made to the Sections of Foundations in Figure E5, if the whole excavated void which is not occupied by the concrete is required to be backfilled, then the cross-section of filling and compaction is the blank areas within the dotted lines, notwithstanding that the contractor may, or may have no option but to, excavate working space. It will be noted from the two diagrams referred to above that, from the practical aspect, the voids as measured to the right of the concrete, are inaccessible for filling. Nevertheless, the measurement rules of the CESMM must apply and it will be for the contractor to allow for working space or other means of backfilling. In such circumstances it would be reasonable to indicate location of the filling in the item description.

Filling and compaction is measured by volume and is given in the Bill of Quantities in m3.

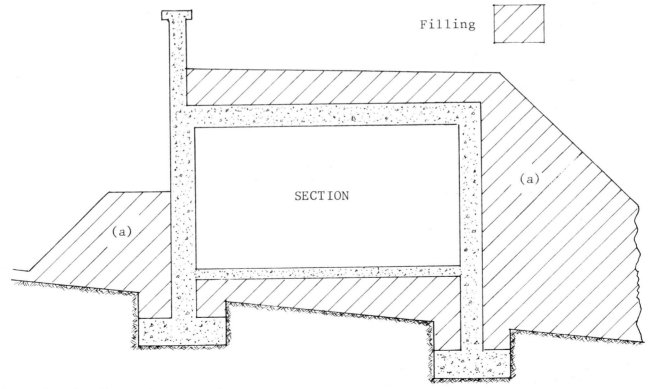

Fig. E6. Filling and compaction to structures.

COMMENTARY

Filling and Compaction (cont.

Embankments

The classification embankments applies not only to those for roads and railways but also those in dam construction.

Embankments are measured by volume and the unit of measurement is m3. The measured volume will include any designed volume for long term settlement.

Any instrumentation to monitor pore pressures and deformation, together with tests during construction would require to be dealt with either as a specified requirement or a p.c. item in the Class A section of the Bill of Quantities.

Specimen item descriptions for the embankment shown in cross-section in the diagram Figure E.6. are given at the side of the diagram.

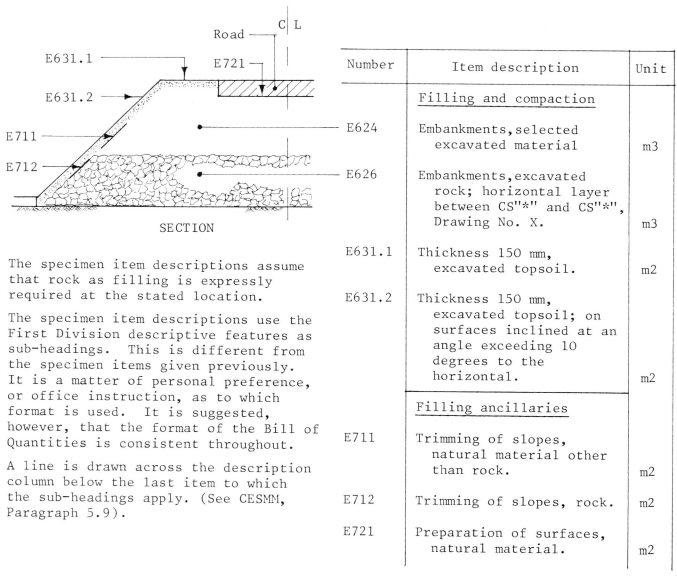

SECTION

Number	Item description	Unit
	Filling and compaction	
E624	Embankments, selected excavated material	m3
E626	Embankments, excavated rock; horizontal layer between CS"*" and CS"*", Drawing No. X.	m3
E631.1	Thickness 150 mm, excavated topsoil.	m2
E631.2	Thickness 150 mm, excavated topsoil; on surfaces inclined at an angle exceeding 10 degrees to the horizontal.	m2
	Filling ancillaries	
E711	Trimming of slopes, natural material other than rock.	m2
E712	Trimming of slopes, rock.	m2
E721	Preparation of surfaces, natural material.	m2

The specimen item descriptions assume that rock as filling is expressly required at the stated location.

The specimen item descriptions use the First Division descriptive features as sub-headings. This is different from the specimen items given previously. It is a matter of personal preference, or office instruction, as to which format is used. It is suggested, however, that the format of the Bill of Quantities is consistent throughout.

A line is drawn across the description column below the last item to which the sub-headings apply. (See CESMM, Paragraph 5.9).

Fig. E7. Embankments.

COMMENTARY

Filling and Compaction (cont.

Filling and Compaction to Stated Thickness

The classification "to stated thickness", applies to filling and compaction of a required thickness, such as a uniform bed or hardcore under a ground slab of a structure, a layer of draining material to form a drainage blanket, and the like.

The classification takes precedence over the other Second Division classifications, it not being required to state whether the filling and compaction so described is in relation to embankments, structures, etc.

Filling and compaction of this classification is measured superficially and the thickness is stated in the item descriptions. The actual thickness having been thus stated it is unnecessary to include the standard phrase "to stated thickness" in the description. When carried out on surfaces inclined at an angle of exceeding 10 degrees to the horizontal the item description must so state. The unit of measurement is m2.

Pitching

The descriptive feature "pitching" in this Class, applies to the facing of earthworks with stones or blocks of hard material, usually placed close together to protect the surface. It should not be confused with pitching as a road base which, if used, would be measurable under Class R.

The work is measured superficially and is given in the Bill in m2. It must be stated in the item description when the work is to surfaces exceeding 10 degrees to the horizontal. All item descriptions for pitching must state the nature and the dimensions and the depth or thickness of the materials used for the work. A specimen of how an item might be worded is as follows:-

Filling and compaction

E657	Pitching, thickness 200 mm, broken limestone size 200 x 150 mm; on surfaces inclined at an angle of exceeding 10 degrees to the horizontal, upstream slopes of dam embankment Drawing X. m2

The sizes in the above description are nominal sizes. It is assumed that the description could be readily related via the Drawing to the specification, where it is assumed tolerances would be given and the method of placing described.

General Filling and Compaction

General filling and compaction is that to make up levels over areas and in situations which would not be appropriately classified under the other Second Division descriptive features. The unit of measurement is m3.

Filling Ancillaries (refer to Table 7.04)

Measured items are given for trimming slopes of filling and the preparation of surfaces of filling. Descriptive features, units of measurement and relevant CESMM Notes are summarised in Table 7.04.

Table 7.05   Landscaping

1st Division - Landscaping

2nd Division			3rd Division
Turfing	m2	State if pegged or wired (Note E28).	Separate items for:-
Hydraulic mulch grass seeding	m2		(i) surfaces not exceeding 10 degrees to the horizontal
Other grass seeding	m2		(ii)surfaces exceeding 10 degrees to the horizontal (Note E28)
Other seeding	m2		
Plants	nr	State species and size	
Shrubs	nr		
Trees	nr		
Hedges	m	State species and size Measure developed length along centre line (Note E29)	State whether single or double row and whether with protective fence, in accordance with the 3rd Division features

COMMENTARY

Landscaping (refer to Table 7.05)

The features of classification for landscaping include turfing, grass seeding, plants, shrubs, trees and hedges. They are set out in Table 7.05 which gives also the units of measurement and a summary of the appropriate CESMM Notes.

Landscaping involves associated work. For example grass seeding in addition to the preparation of the seed bed, may require the application of fertiliser, watering, weeding, mowing, etc. Item descriptions or preamble make clear the intended item coverage where it would otherwise not be clear.

Turfing

Turfing is measured by area and is given in m2. Separate items are given for that laid on surfaces not exceeding 10 degrees to the horizontal and for that on surfaces exceeding 10 degrees to the horizontal. Item descriptions state where turfing is to be pegged or wired. They make distinction between turfing carried out with imported turves and those obtained from the Site.

Hydraulic Mulch and Other Grass Seeding

Grass seeding is given in m2. That on surfaces not exceeding 10 degrees to horizontal is given separately from that on surfaces exceeding 10 degrees to horizontal. The specification for hydraulic mulch grass seeding will usually require that this type of seeding is carried out in accordance with the recommendations issued for the particular process and item descriptions make reference to the appropriate clauses.

COMMENTARY

Landscaping (cont.

Plants, Shrubs, Trees and Hedges

The specification or drawing will usually indicate the botanical names of the species and these are the names stated in the item descriptions. Preamble or additional description should make clear the associated work such as pits, staking, ties, etc., which is intended to be included in the items.

Unless other particular feature is required, the size of shrubs is given in terms of height above ground.

Trees, in addition to their botanical names, will usually be described as "half-standard", "standard", "heavy standard", "semi mature", etc., and each of these types will be defined in the specification in terms of nominal height above ground, height to the first branch and diameter at a given height above ground.

Hedges are measured in linear metres, on their centre line, to their developed length. This means that the measured length will be from the first to the last hedge plant in the row, to which will be added half the distance between the hedge plants for each end. Separate items are given for hedges with single rows of plants, those with double rows of plants and those with protective fences.

Specimen Item Descriptions for Landscaping

The specimen item descriptions in the adjoining column assume the Specification clauses describe:-

the turfing and whether or not the turves are to be obtained from the Site,

the seeding and associated work,

the shrubs and associated work and how height is measured,

the trees and associated work and the size of a "standard" tree,

the hedge plants and associated work and how height is measured.

Number	Item description	Unit
	Landscaping	
E811	Turfing, to surfaces inclined at an angle of not exceeding 10 degrees to the horizontal; as Specification clause "*".	m2
E832	Grass seeding, to surfaces inclined at an angle exceeding 10 degrees to the horizontal; as Specification clause "*".	m2
E860	Shrubs, Buxus sempervirens, 600 mm high; as Specification clause "*".	nr
E870	Trees, Quercus robur, standard; as Specification clause "*".	nr
E881	Hedges, Fagus sylvatica, 600 mm high, single row; as Specification clause "*"	m

NOTE: "Buxus sempervirens" = Common Box.
         "Quercus robur" = Common Oak.
         "Fagus sylvatica" = Common Beech.

EXAMPLE No. EE.1

Measured Example

The Example of "taking off" which follows illustrates conventions and the application of the provisions of the CESMM to the measurement of the earthworks for the Cutting and Embankment shown on Drawing No. E/D/1. The cross-sections in the Example are of fairly regular shape and the ordinary trapezoidal rule is used for the calculation of areas and volumes. The Example does not include the measurement of the road.

Initial Calculations for Dimensions

For the measurement of a cutting and filling operation of the kind in the Example, numerous waste calculations need to be made before a meaningful start can be made to set down the dimensions in the dimension column of the sheets. It would cause lengthy written work on the dimension sheets if measurements were taken and entered for small sections of the operations. It is best to make a schedule of initial waste calculations, setting out the calculated depths, slope lengths and the like, in a form convenient for reference and to proceed on the dimension sheets when results from the schedules can be collected and formulated into as few as possible entries against the items.

Confusion will result if the method of setting down the initial calculations does not allow an easy identification of the depth or other measurement of a particular part of the work.

In working the example the initial calculations are made and set out in the form shown on Plate EP.1. Set out in this manner, particular measurements are easily located as the "taking off" proceeds. If the schedule is referred to when measurement work for the final account is proceeding on the site it will also be simple to check the dimensions which were used to calculate the Quantities.

The sheet for the schedule is ordinary analysis paper, ruled with the requisite number of vertical columns. The first four steps in the procedure of preparation are as follows:-

1. Enter appropriate cross-section reference.

2. Enter road level given at the centre line of the cross-section

3. Calculate the road levels at the sides, i.e. take $2\frac{1}{2}\%$ of half the constant width (14 m divided by 2). Add resultant to the centre level for the right. Deduct resultant from centre level for the left. Enter on schedule.

4. Deduct existing ground levels from road levels. The resultant "D" = depth. A negative resultant indicates cut. A positive resultant indicates fill.

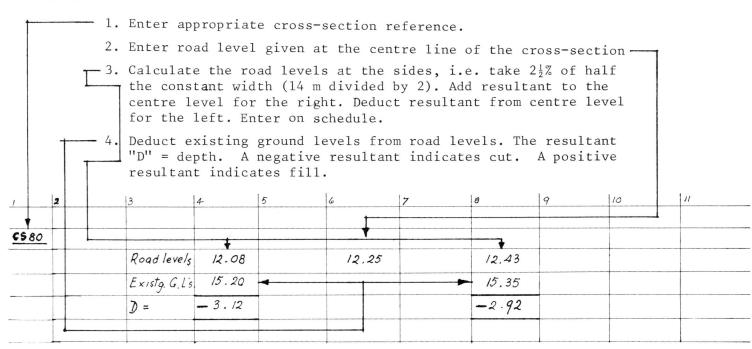

1	2	3	4	5	6	7	8	9	10	11
CS 80										
		Road levels	12.08		12.25		12.43			
		Existg. G.L's.	15.20				15.35			
		D =	− 3.12				−2.92			

EXAMPLE No. EE.1

Measured Example (cont.

Initial Calculations for Dimensions

Continuing the explanation of the procedure of preparation of the schedule of Initial Calculations, the further steps are as follows:-

5. Calculate SB = spread of banks, by multiplying "D" by the Cotangent of the angle at the foot of the banks (Cotangent of 30 degrees = 1.73205). Enter the figures across the vertical columns 3 & 4 and 8 & 9.

6. Enter "CW" - constant width. Thicken lines to form panels to distinguish horizontal measurements from those for depths, slope lengths, etc.

7. Add spread of banks to "CW". Enter overall width. (Top for cuttings, bottom for embankments).

8. Add 6 & 7 above together and divide by 2 to calculate mean width. Enter mean width.

9. Add together "D" on left and "D" on right and divide by 2, to calculate "AVD" = average depth. Enter average depth.

10. Multiply ("AVD" by the Cosecant of 30°) to calculate average slope length. (Cosecant of 30 degrees = 2.0000). Draw sloping lines and enter slope lengths.

	2	3	4	5	6	7	8	9	10	11
				AVS		AVS			WIDTHS	
		SB=	5.40 (5)					5.06 (5) =	24.46 (7)	top
CS80	(1)	CW=			14.00 (6)				14.00 (6)	bottom
		Road levels	12.08 (3)		12.25 (2)	6.04 (10)	12.43 (3)		(19.23)(8)	mean
		Existg. GL's	15.20 (4)	6.04 (10)		6.04 (10)	15.35 (4)			
		D =	3.12 (4)		3.02		2.92 (4)			
					AVD (9)					

When the schedule is completed to the extent shown above all waste calculations for CS80 have been entered. The same procedure is then followed for CS60 and so on until all the wastes for all cross-sections have been entered. The distances between cross-sections are then entered in the vertical column 2 and the position of the zero contour is calculated all as shown on Plate EP.1. All is then organised for recording the items and dimensions on the dimension sheets. It should be remembered that in this case the finished sizes are recorded on the schedule.

Approach to the Taking of Dimensions

From an inspection of the Drawing, an initial break down on the work into the following parts or elements is considered convenient for the purpose of measurement:- (i) excavation, (ii) filling, and (iii) work to surfaces. The Example deals with the work in that order, completing the measurements of the items within each part or element before proceeding to the next. It being considered that this is a methodical sequence of measurement.

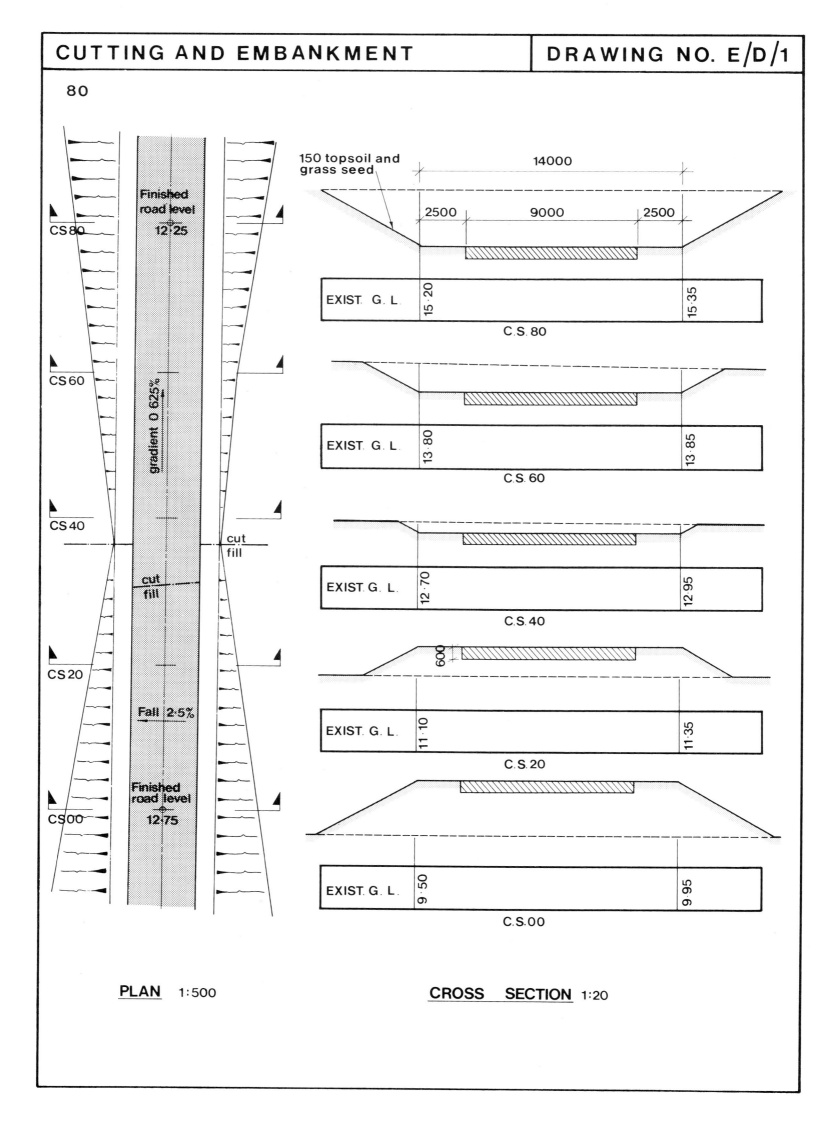

CUTTING AND EMBANKMENT

DRAWING NO. E/D/1

PLAN 1:500

CROSS SECTION 1:20

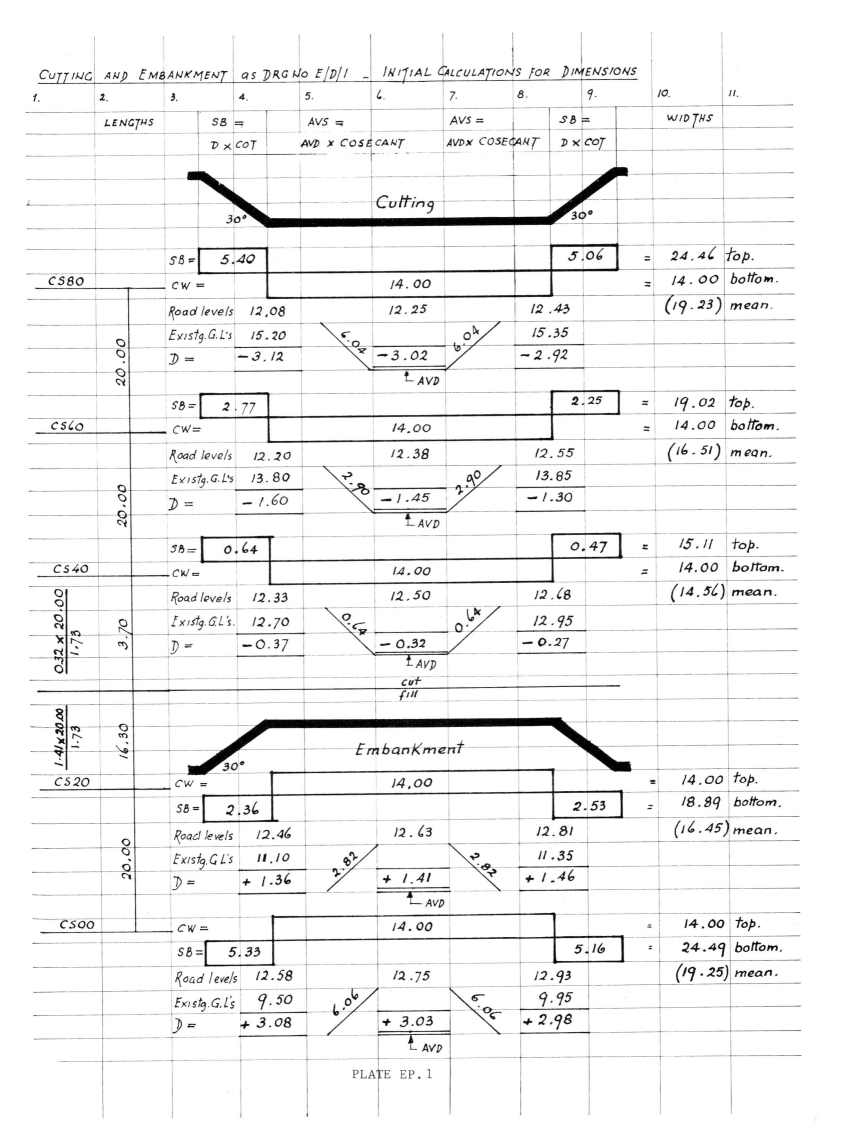

CUTTING AND EMBANKMENT as DRG No E/D/1 — INITIAL CALCULATIONS FOR DIMENSIONS

1.	2.	3.	4.	5.	6.	7.	8.	9.	10.	11.
	LENGTHS		SB = D × COT	AVS = AVD × COSECANT		AVS = AVD × COSECANT		SB = D × COT	WIDTHS	

Cutting 30°   30°

SB =	5.40						5.06	= 24.46 top.
CS80	CW =			14.00				= 14.00 bottom.
	Road levels	12.08		12.25		12.43		(19.23) mean.
	Existg. G.L's	15.20	6.04		6.04	15.35		
	D =	−3.12		−3.02		−2.92		
				↑ AVD				

20.00

SB =	2.77					2.25	= 19.02 top.
CS60 CW =			14.00				= 14.00 bottom.
Road levels	12.20		12.38		12.55		(16.51) mean.
Existg. G.L's	13.80	2.90		2.90	13.85		
D =	−1.60		−1.45		−1.30		
			↑ AVD				

20.00

SB =	0.64					0.47	= 15.11 top.
CS40 CW =			14.00				= 14.00 bottom.
Road levels	12.33		12.50		12.68		(14.56) mean.
Existg. G.L's	12.70	0.64		0.64	12.95		
D =	−0.37		−0.32		−0.27		
			↑ AVD				

$\dfrac{0.32 \times 20.00}{1.73}$ 3.70

cut / fill

$\dfrac{1.41 \times 20.00}{1.73}$ 16.30

Embankment 30°

CS20 CW =			14.00				= 14.00 top.
SB =	2.36					2.53	= 18.89 bottom.
Road levels	12.46		12.63		12.81		(16.45) mean.
Existg. G.L's	11.10	2.82		2.82	11.35		
D =	+1.36		+1.41		+1.46		
			↑ AVD				

20.00

CS00 CW =			14.00				= 14.00 top.
SB =	5.33					5.16	= 24.49 bottom.
Road levels	12.58		12.75		12.93		(19.25) mean.
Existg. G.L's	9.50	6.06		6.06	9.95		
D =	+3.08		+3.03		+2.98		
			↑ AVD				

PLATE EP. 1

EARTHWKS - CUTTS + EMBANKS. DRG. E/D/1

EXAMPLE EE.1

<u>Strip topsoil</u>

CS 80   ½/ 24.46 = 12.23
CS 60   19.02 = 19.02
CS 40   ½/ 15.11 = <u>7.56</u>
                38.81

CS 40   ½/ 15.11 = 7.56
O   ½/ 14.00 = <u>7.00</u>
                14.56

O   ½/ 14.00 = 7.00
CS 20   ½/ 18.89 = <u>9.45</u>
                16.45

CS 20   ½/ 18.89 = 9.45
CS 00   ½/ 24.50 = <u>12.25</u>
                21.70

20.00			Excavn of cutts., soil for disposal; Excavd. Surf. 150 mm. below Orig. Surf.
38.81			
0.15	116.43		E 220.1
3.70			
14.56			
0.15	8.08		(CS40 - 0
16.30			
16.45			
0.15	40.22		(CS 0 - 20
20.00			
21.70			
0.15	65.10		(CS 20 - 00
2/ 43.70			2/20.00 = 40.00
0.04			3.70
0.15	0.52		(overlap top cuttings 43.70
	230.35		

(1)

(1.)

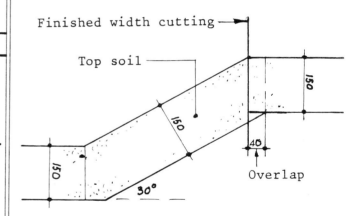

Cutts.

Bulk dig

$150 \times 1.1547 = 173$

         150

         23

$23 \times 1.732 = 40$

	19.23	16.51	14.56	14.56
2/.04 =	.08	.08	.08	.08
	19.31	16.59	14.64	14.64

½/ 20.00		
19.31		Excavn. of cutts, matl. for disposal, Comm. Surf. u/s topsoil, Excavd Surf 450 mm above Fin. Surf (CS 80
3.02	583.16	
20.00		E 240.1
16.59		
1.45	481.11	(CS 60
½/ 20.00		
14.64		
0.32	46.85	(CS 40
½/ 3.70		
14.32		
0.32	8.48	(CS 40 - 0
	1119.60	

     14.64

     14.00

2) 28.64

     14.32

Addn. depth road

$0.32 + 0.45 = 0.77$

$\dfrac{0.77 \times 20.00}{1.73} = 8.90$

         3.70

         5.20

2/ 20.00		
9.00		Excavn. of cutts., matl. for disposal; Comm. Surf. 450mm above Fin. Surf
0.45	162.00	
3.70		E 240.2
9.00		
0.45	14.99	
½/ 5.20		
9.00		
0.45	10.53	
	187.52	

(2)

An adjustment in waste is made to the finished width to bring it to the width of the bulk excavation. This is necessary because of the difference indicated in the diagram in the Commentary to the Dimensions in Column (1).

The depth of excavation is measured to the formation to receive the top soil of the verges at the sides of the roads. It is the same as the finished depth. The 150 mm top soil removed compensating for the 150 mm to be placed at the bottom.

The dimensions are set down, length x average width x average depth. The last two dimensions in each set being those for the areas of the cross-section. The weighting being, 1 for the first and last cross-section and 2 for the intermediate cross-sections.

It will be noted that the descriptions state an Excavated Surface or a Commencing Surface because the excavation starts or finishes at other than the Commencing Surface or Final Surface, respectively.

The description added to the standard descriptions generates a dotted on suffix number.

The last item in the adjoining column is the additional depth to accommodate the road, as shown in the following diagram. The zero contour line changes position because of the additional depth.

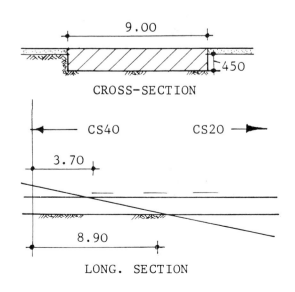

CROSS-SECTION

LONG. SECTION

EARTHWKS - CUTTS. & EMBANKS. DRG. E/D/1

2/0.04	16.45 19.25
	0.08 0.08
	16.37 19.17

Embanks.

½/	16.30	
	15.19	
	1.41	174.56

Fillg and compactg. to embanks. sel. excavd matl

E624   (CS 0.10

	16.37
	14.00
	2)30.37	(CS 20
	15.19

½/	20.00	
	16.37	
	1.41	230.82

½/	20.00	
	19.17	
	3.03	580.85
		986.23

Ddt. Excavn. of cutts. matl. for disposal, a.b.

E240.1   (CS.00

+

Add. Excavn. of cutts. matl. for re-use; Comm. Surf. u/s. topsoil Excavd Surf 450mm above Fin. Surf E230.1

Sinkg. for road

  20.00
   8.90
  11.10

½/	5.20	
	9.00	
	0.45	10.53
	11.10	
	9.00	
	0.45	44.96
	20.00	
	9.00	
	0.45	81.00
		136.49

Ddt. Fillg. and compactg. to embanks, a.b. E624

+

Ddt. Excavn of cutts., matl. for re-use, a.b. E230.1

+

Add. Excavn of cutts, matl for disposal, a.b. E240.1

(3)

The waste calculations adjust the finished width of the embankment to the required bulk filling width. See diagram below.

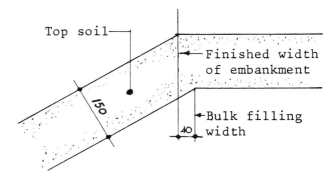

The depth of the bulk filling and compaction is the same as the finished depth, the 150 mm top soil removed below the embankment compensating for the 150 mm to be placed at the top.

It is assumed that it will be possible to select sufficient suitable material for filling from the excavations, also that the filling material will involve only one compaction required.

Having previously measured all the excavation as in material for disposal, the excavation which will provide the filling is re-classified as for re-use by the "Add" and "Deduct".

The abbreviation "a.b." means "as before". In this case the Code number follows the abbreviation and the intention is that the deduction shall be made from the item with that Code number as previously described. This avoids writing out details which can be obtained from the item referenced.

The first series of dimensions in the adjoining column are taken to the underside of the top soil to the verges. The sub-base of the road is sunk below this. Consequently, a deduction is made for the sinking. The reduction in the amount of filling means less material for re-use is required and a volume of excavation equal to the filling is re-classified.

<u>Bottom of cutts to</u>
<u>receive sub-base.</u>

		2/20.00 = 40.00
		3.70
		5.20
		48.90

48.90		<u>Excavn. ancills., prepn.</u>
9.00	440.10	of surfs. nat. matl.
		E 521.1

<u>Sides of sub-base.</u>

2/ 48.90		Excavn. ancills., prepn.
0.45	44.01	of surfs., nat. matl.;
		vertical
		E 521.2

<u>Surfs. embanks to</u>
<u>receive sub-base.</u>

		80.00
		48.90
		31.10

31.10		Fillg. ancills., prepn.
9.00	279.90	of surfs. nat. matl.
		E 721.1

<u>Sides of sub-base</u>

2/ 31.10		Fillg. ancills., prepn.
0.45	27.99	of surfs., nat. matl;
		vertical
		E 721.2

(4)

The preparation of surfaces might conventionally be measured when the dimensions of the road were being taken. The actual detail of the Permanent Works in contact with the earthworks would be more apparent from the roadwork drawings and dimensions could be generated from the dimensions of the road.

Preparation of surfaces is measured to both excavated and filled surfaces. It is measured to both horizontal and vertical surfaces which are to receive Permanent Works. It is possible that at the edges, the filling would be overfilled and cut back to the vertical surface. These edges are classified as the preparation of filled surfaces and not as excavated surface as the method of preparation might suggest.

It is considered that there will be a difference in cost between the preparation of vertical surfaces of excavation and the preparation of horizontal surfaces. In accordance with Paragraph 5.10 of the CESMM, the word "vertical" has been added to the standard description where the surfaces to be prepared are vertical. This makes it necessary to dot on a suffix number to the Code number of both the standard description and the amplified description.

It is assumed that the requirements of the specification are merely to prepare the formation and that a surface dressing or other work to the formation is not called for. Had this not been the case, the surface dressing or other work would be given as separate items, or if preferred, could be included with the preparation item by adding the description of the work to the standard description.

EARTHWKS. - CUTTS. & EMBANKS. DRG. E/D/1

<u>Verges & top cutts</u>

4/20.00 = <u>80.00</u>

2/	80.00	
	2.50	400.00
2/	43.70	
	0.04	3.50
		403.50

{ Fillg. and compactg., thickn. 150 mm, excavd. soil.

E 631.1 (top banks

&

<u>Ddt.</u> Excavn. of cutts., soil for disposal, a.b. E 220.1
Cub x 0.15 = 60.53 m3

&

<u>Add.</u> Excavn. of cutts., soil for re-use, Excavd. Surf. 150 mm below Orig. Surf.
Cub x 0.15 = 60.53 m3

E 210.1

&

<u>Add.</u> Landscapg., grass seedg., surfs. n.e. 10 degrees; Spec. clause X.

E 831.1

(5)

COMMENTARY

The first set of dimensions in the adjoining column are for the verges at the sides of the road. The second set covers the replacement of top soil to the overlap at the top of the banks to the cutting. The dimensions for the overlap are obtained from those for excavation.

It is necessary to dot on a suffix number to the Code for the top-soiling as explained in the Commentary to Column (7) of the Dimensions.

When measuring the excavation of top soil, it was classified as in material for disposal. It is necessary to re-classify, as for re-use, sufficient volume for the filling. This is done by 'Deduct" and "Add".

The filling with topsoil is measured superficially, whilst the excavation is measured cube. Rather than write out fresh dimensions for the excavation items, the instructions to cube the areas attached to the items by the depth of the topsoil is given in the description column for each item requiring to be cubed.

The landscaping item has a suffix number dotted on to the Code to indicate that description additional to the standard features has been included.

EARTHWKS - CUTTS & EMBANKS. DRG. E/D/1

Slopes to cutts. CS 80-0

2/½/	20.00	
	6.04	120.80
2/	20.00	
	2.90	116.00
2/½/	20.00	
	0.64	12.80
2/½/	3.70	
	0.64	2.37
		251.97

{ Excavn. ancills., trimmg of slopes, nat. matl.

E 511

&

{ Fillg. & compactg. thickn 150 mm, excavd. soil, surfs. ov. 10 degrees.

E 631.2

&

{ Landscapg., grass seedg. surfs. ov. 10 degrees; S.pec clause X

E 832.1

&

{ _Ddt. Excavn. of cutts., soil for disposal, a.b. E 220.1_

Cub. x 0.15 = 37.80 m³

&

{ _Add. Excavn. of cutts., soil for re-use a.b. E 210.1_

Cub. x 0.15 = 37.80 m³

(6)

The slope lengths are taken from the Initial Calculation Sheet. As there are two sides to the cutting the dimensions are timesed by 2/.

The procedure for entering the dimensions for the areas of the slopes uses the slope length at the cross-sections multiplied by the length between the cross-sections with a weighting of ½/ for the ends and 1/ for the intermediates and is illustrated diagramatically below.

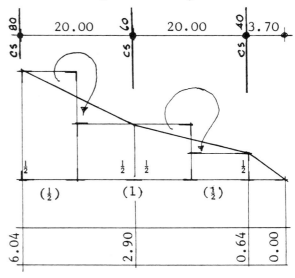

If preferred the slope lengths at the cross-sections could be averaged in waste and the dimensions entered as follows:-

2/	20.00	)		6.04
	4.47	)		2.90
2/		)	2)	8.94
	20.00	)		4.47
	1.77	)		
2/		)		2.90
	3.70	)		0.64
	0.32	)	2)	3.54
		)		1.77
				0.64
				0.00
			2)	0.64
				0.32

For comments on topsoiling and also excavation adjustment : See the Commentary to Columns (5) and (7) of the Dimensions.

EARTHWKS - CUTTS & EMBANKS. DRG E/D/1

Slopes to embanks

2/½/	16.30	
	2.82	45.97
2/½/	20.00	
	2.82	56.40
2/½/	20.00	
	6.06	121.20
		223.57

Fillg. ancills., trimmg. of slopes, nat. matl.
E 711

&

Fillg & compactg. thickn 150 mm, excavd. soil, surfs. ov. 10 degrees
E 631.2

&

Landscapg. grass seedg. surfs ov. 10 degrees
a.b. E 832.1

&

Ddt. Excavn of cutts. soil for disposal
a.b. E 220.1
Cub. x 0.15 = 33.54 m3

&

Add. Excavn of cutts. soil for re-use
a.b. E 210.1
Cub x 0.15 = 33.54 m3

The dimensions at the Sections are entered, following the same procedure as explained in the Commentary to Column (6) of the Dimensions.

The slopes of the bulk filling are measured for trimming to receive the uniform layer of topsoil.

Topsoiling is measured according to the classification "stated thickness". Because the actual thickness is stated there could be several items with different thicknesses which would attract the same standard Code number. Different suffix numbers are dotted on to the Code number for each different thickness or added description.

When measuring the excavation of topsoil, it was classified as in material for disposal. It is necessary to re-classify, as for re-use, sufficient volume for the filling and the last two items do this.

NOTE: The dimensions in this Example have been squared for the purpose of preparing the Specimen Abstract at the end of Chapter 1. In practice dimensions would be lined through when they are abstracted. For clarity this has not been done in the Example.

8 In Situ Concrete, Concrete Ancillaries and Precast Concrete—CESMM Classes: F, G and H

Classes F, G and H of the CESMM cover concrete structures and general concrete work with certain exceptions.

Class F relates exclusively to in situ concrete. Formwork and reinforcement for in situ concrete, post-tensioned prestressing and sundry ancillaries for in situ concrete are included in Class G. Precast concrete units are covered by Class H. The foregoing is an outline of the work covered by the three Classes. References are made subsequently to the work specifically included in and excluded from each Class. The text of this Chapter deals with each Class individually in alphabetical order of the Class references.

IN SITU CONCRETE - CESMM CLASS F

No specific "Includes" are listed at the head of the Class F classification table in the CESMM. The Class covers in situ components for structures and other forms of in situ concrete construction with the exception of those specifically excluded. For work excluded from the Class refer to the "Excludes" at the head of the Class F classification table in the CESMM.

The rules in Class F require items for the provision of concrete to be given separately from those for the placing of the concrete. Both provision and placing are measured by volume and are given in the Bill of Quantities in m3.

Rules regarding deductions from the volumes of concrete and the volumes to be ignored are outlined in Table 8.01. These rules apply to both the provision and placing of concrete and are set out in Notes F10 and F11 of the CESMM.

The diagram in Figure F1 illustrates the terms internal and external splays used in Notes F10 and F11 of the CESMM.

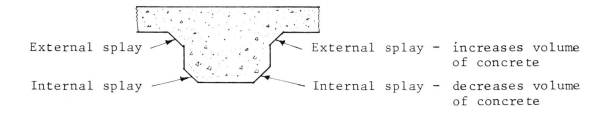

External splay — increases volume of concrete

Internal splay — decreases volume of concrete

Fig. F1. Internal and external splays.

Table 8.01   Concrete Generally

No deduction from volume of concrete for (a) reinforcement and metal sections
(b) prestressing components (c) cast in components less than 0.1 m3 (d) rebates,
fillets or internal splays less than 0.005 m2 cross-sectional area (e) pockets
and holes measured as large or small voids in accordance with Note G7 (Note F10).

The volume of concrete in nibs and external splays less than 0.005 m2 cross-
sectional area is not measured.   (Note F11).

Table 8.02   Provision of Concrete

Generally - Classify and describe in accordance with CP 110: Part 1: 1972,
Section 6.1 (Note F1). Designed mixes are those where the mix proportions
are selected by the Contractor. Prescribed mixes are those where the mix
proportions are selected by the Engineer.

Ordinary Structural Concrete

1st Division	2nd and 3rd Divisions	
Designed mix for ordinary structural concrete m3	State the Grade of concrete (as 2nd Division), the size of aggregate (as 3rd Division) and additionally the type of cement and the minimum cement content in kg per m3.	N.B. Can only contain cement to BS 12, BS 146, or to BS 4027 and must not contain special additives.
Prescribed mix for ordinary structural concrete taken from Table 50 of CP 110; Part 1. m3	State the Grade of concrete (as 2nd Division), the size of aggregate (as 3rd Division) and additionally the type of cement.	The Grade of concrete is the same as the 'characteristic strength' defined in CP 110. i.e. the 28 day crush-ing strength in N/mm2.
Prescribed mix for ordinary structural concrete not taken from Table 50 of CP 110; Part 1. m3	State the Grade of concrete (as 2nd Division), the size of aggregate (as 3rd Division) and additionally the type of cement, the minimum cement content in kg per m3, the proportions of cement to fine aggregate to coarse aggregate, the type of aggregate and the required workability.	

Special Structural Concrete

1st Division	2nd and 3rd Divisions
Designed mix for special structural concrete m3	These mixes should be fully described in the Specification and item descriptions should refer to the appropriate Specification clause.
Prescribed mix for special structural concrete m3	

COMMENTARY

Provision of Concrete (Refer to Table 8.02)

The terms used to describe the concrete are "ordinary structural concrete" or "special structural concrete" and each description will be classified as either a "designed mix" or a "prescribed mix" according to the requirement. The details which need to be stated in an item description for the provision of concrete are the same as the requirements to be specified as provided in subsection 6.1.3 of C.P. 110: Part 1, 1972. An outline of the details to be stated in the item descriptions are given in Table 8.02.

In practice, rather than list all the details required by the CESMM in the items, it will usually be found convenient to describe the various ordinary and special concrete mixes in the Specification and to tabulate them and allocate each particular mix a reference. The item descriptions will then need only to quote the Specification reference. For example:-

Provision of concrete

Type D20 as Specification clause 'X'.   m3

The code number for the above item would follow the CESMM convention. For example, if Type D20 concrete is a designed mix for ordinary structural concrete, grade 20, with 20 mm aggregate, it would be coded F133.

Placing of Concrete Generally (Refer to Table 8.03)

Item descriptions for placing concrete state whether the concrete is mass concrete, reinforced concrete or prestressed concrete. Prestressed concrete which is also reinforced is classed as prestressed.

The descriptions will also state the element in which the concrete is placed in accordance with the Second Division descriptive features and, except for the classification "other concrete forms", will give the dimension ranges of the components as provided in the Third Division of the classification table, Class F of the CESMM. It will be noted from Table 8.03 that actual dimensions require to be stated, in place of a dimension range, where only one thickness or cross-sectional area is included in one item. For the details given at Third Division level of the classification "other concrete forms", see Table 8.03.

It is important to implement Note F2 of the CESMM and state in the item descriptions the location of concrete components where special characteristics may affect the method and rate of placing.

Blinding

The thickness used for the classification of blinding is the minimum thickness. See Note F4 of the CESMM. The Note should be implemented in relation to blinding in a particular location and not to blinding as a whole. For example, if say 75 mm thick blinding occurs in one location and say 200 mm thick in another, each would be itemised and given according to its particular thickness range. Interconnected blinding of varying thicknesses would be classified according to the minimum thickness.

Bases, Footings and Ground Slabs

The placing of concrete in the several elements bases, footings and ground slabs are grouped in a single descriptive feature. The volume of all three elements of

Table 8.03 Placing of Concrete

Generally - State the location of the members in the Works where special characteristics may affect method and rate of placing concrete (Note F2). Note F2 is worded "may be stated" but it is considered in the interest of good practice to read this as "shall be stated". Among the factors to be considered when implementing this Note are height above and depth below ground, plan position, access, rate of pour, density of reinforcement and curing requirements.

1st Division

    Mass

    Reinforced

    Prestressed   Includes prestressed which is also reinforced (Note F3)

2nd Division			3rd Division	
Blinding	m3		State 3rd Division thickness range, except where only one thickness in one item. Use minimum thickness for classification of blinding. (Note F4).	State actual thickness where only one thickness included in one item. (Paragraph 5.14)
Bases, footing and ground slabs	m3	Measure beams attached to ground slabs as part of slab (Note F8)	State 3rd Division thickness range, except where only one thickness in one item. Thickness for classification of ground slabs should exclude thickness of attached beams.	
Suspended slabs	m3	Measure beams attached to suspended slabs as part of slab. (Note F8) Suspended slabs less than one metre wide shall be classed as beams.(Note F9)	State 3rd Division thickness range, except where only one thickness in one item. Thickness classification for suspended slabs shall exclude additional thickness of attached beams. (Note F9).	
Walls	m3	Measure columns and piers attached to walls as part of wall. (Note F5). Walls less than one metre long shall be classed as columns (Note F9).	State 3rd Division thickness range, except where only one thickness in one item. Thickness classification for walls shall exclude additional thickness of attached columns and piers (Note F9)	

Table 8.03 Placing of Concrete (cont.

2nd Division			3rd Division
Columns and piers	m3	Not applicable to columns and piers attached to walls. See Walls.	State the 3rd Division cross-sectional area range, except where only one cross-sectional area in one item. State actual cross-sectional area where only one included in one item
Beams	m3	Not applicable to beams attached to slabs. See Slabs	State the 3rd Division cross-sectional area, except where only one cross-sectional area in one item.
Casing to metal sections	m3	Not applicable to casing of metal sections integral with walls and slabs. See Walls and Slabs	State actual cross-sectional area where only one included in one item. Special beam sections – Beams are classed as special beam sections where their cross-section profiles are rectangular (or approximately) over less than 4/5 of their length or where of box or other composite section. State cross-sectional dimensions or state type or mark number for which dimensions are given on a Drawing. (Note F6).
Other concrete forms	m3		State (a) principal dimensions of component, or (b) type or mark number where principal dimensions are given on the Drawings, or (c) state location when principal dimensions are given on the Drawings. (Note F7).

COMMENTARY

Placing of Concrete (cont.

Bases, Footings and Ground Slabs (cont.

the same thickness range are added together in one item, unless further itemisation is considered advisable in accordance with Note F2 of the CESMM. (See "Generally", Table 8.03).

Beams integral with ground slabs are measured as part of the slab. Although not specifically stated in the CESMM that it shall be so, the classification thickness of the ground slab element is taken as that of the slab.

Suspended Slabs

The volume measured for the placing of concrete in a suspended slab will include the volume of beams integral with the slab. It will also include the volume of any other projections integral with the slab which are considered to involve complexity of placing no greater than that of the slab. Projections considered to involve complexity of placing greater than that of the slab would be appropriately classified and be given separately. For notes on classification thickness and slabs less than one metre in width see those against the feature in Table 8.03.

COMMENTARY

Placing of Concrete (cont.

Walls

The volume measured for the placing of concrete walls will include the volume of any columns and piers integral with the wall. The thickness classification will be that of the wall excluding the thickness of any attached columns or piers. Walls less than one metre long are classed as columns. Item descriptions for walls of special profile will usually include a drawing reference.

Columns and Piers

The classification "columns and piers" is appropriate for independent members. It is not applicable to columns and piers attached to walls.

Column caps may be classed as columns or as "other concrete forms". In either case they are identified as column caps in the item descriptions. Alternatively, column caps may be measured as part of the slab to which they are attached.

Beams

The classification "beams" does not apply to beams integral with slabs.

The classification "special beam section" is defined in Note F6 of the CESMM. See note under 3rd Division in Table 8.03.

Casing to Metal Sections

The sections cased are identified in the item descriptions, i.e. stanchions, beams, grillages, etc. The classification does not apply to the casing of metal sections which is integral with walls or slabs.

Other Concrete Forms

Particular identifying details are given in the item descriptions for components classified as "other concrete forms". These are set out in Note F7 of the CESMM. See note under 3rd Division in Table 8.03.

The classification is appropriate for use where (i) a component would not be correctly described by a standard descriptive feature, or where (ii) a combination of work items form a member which it would be more practical to treat as composite.

Tapering and Sloping Members

Tapering members whose thickness or cross-sectional area extends into more than one thickness or cross-sectional area range are divided into separate items for each standard thickness or cross-sectional area range, respectively.

Where there is doubt as to whether a sloping member should be classed a beam or a column, the classification "other concrete forms" is appropriate, or it can be classed as either a column or a beam according to judgement. In the latter case it may be considered helpful to give the location of the member or state a drawing reference.

Concrete ancillaries in Class G covers formwork and reinforcement for in situ concrete, to the extent that they are not listed in the "Excludes" at the head of the classification table. The Class includes joints in in situ concrete, post tensioned prestressing and accessories for in situ concrete. It excludes post tensioned prestressing which is included in Class H. Reference to work included in Class G is made in the "Excludes" at the head of the classification tables in Classes M, N, O and W.

Table 8.04   Formwork

Generally   Formwork will be deemed to be to plane areas exceeding 1.22 m wide unless otherwise stated (Note G5)

Formwork is not measured to blinding concrete (Note G2)

1st Division

Formwork: Measure for all final surfaces of in situ concrete requiring support during casting, unless otherwise stated (Note G1)

rough finish

fair finish   Measure to side surfaces of in situ concrete where cast within excavated volumes and not expressly required to be cast against excavated surfaces, including back sloping surfaces not exceeding 45 degrees to the vertical (Note G1)

other stated finish

stated surface features   Measure to upper surfaces exceeding 15 degrees to the horizontal and state that it is to upper surfaces except where not exceeding 10 degrees to the vertical (Note G1)

Measure to temporary surfaces where expressly required (Note G2)

2nd Division		3rd Division
Plane horizontal	85 - 90 degrees to the vertical (Note G3)	State width ranges:
Plane sloping	10 - 85 degrees to the vertical (Note G3)	not exceeding 0.1 m m
Plane battered	0 - 10 degrees to the vertical (Note G3)	0.1 - 0.2 m m
Plane vertical	0 degrees to the vertical (Note G3)	0.2 - 0.4 m m2
		0.4 - 1.22 m m2
Curved to one radius in one plane	State radius (Note G6a)	exceeding 1.22 m m2
Other curved to stated radii m2	One radius in two planes (spherical), state radius (Note G6b)	
	Varying radius (conical) state maximum and minimum radii (Note G6c)	

N.B.   The formed surfaces of rebates, grooves and fillets shall be classed as plane surfaces (Note G5)

Fig. F2. Plane formwork, classifications and angles of inclination in accordance
          with Note G3 of the CESMM

Table 8.04 Formwork (cont.

1st Division	2nd Division	3rd Division
Formwork: rough finish fair finish other stated finish stated surface features	For voids nr	Classify as to large or small voids: Large = 0.1 - 0.5 m2 area, or where circular 0.35 - 0.7 m diameter (Note G7) Small = not exceeding 0.1m2 area, or where circular not exceeding 0.35 m diameter (Note G7) Measure depth perpendicularly to the adjacent surface of concrete (Note G7). Classify in accordance with depth ranges. State actual depth where exceeding 2 m and also where only one depth in one item Measure in detail when larger than "large voids" (Note G8) No deduction from other formwork for that obscured by the forms for "large" and "small" voids (Note G9)

	2nd Division		3rd Division
	For concrete components of constant cross-section m	This classification may be used to measure the formwork by length as one item instead of by area in items for the separate surfaces where the formwork is for components of constant cross-section. State the principal cross-section dimensions of the component and its mark number, location or other unique identifying feature (Note G4)	State whether:- Beams Columns or Walls Describe "other members"

COMMENTARY

Formwork (Refer to Table 8.04)

Formwork is measured to the surfaces stated in Note G1 of the CESMM. See the 1st Division panel of Table 8.04. All formwork is classified at First Division level according to the finish or features it is to produce on the surface of the concrete. Commentary follows which is particular to the further levels of classification.

Formwork to Plane Surfaces

Item descriptions for formwork to the plane surfaces of concrete describe the formwork as horizontal, sloping, battered or vertical, each indicating defined angles of inclination as noted in Table 8.04. See also Figure F2. Descriptions state also the appropriate Third Division width range for formwork not exceeding 1.22 m wide. Quantities for formwork in the ranges not exceeding 0.2 m wide are given in linear metres. Those for formwork in the ranges exceeding 0.2 m wide are given in m2. It is not necessary to state the width when it exceeds 1.22 m wide. (See note against "Generally" in the first panel of Table 8.04). Width is taken as the least of the two dimensions of the formwork.

COMMENTARY

Formwork (cont.

Curved Formwork

"Formwork curved to one radius in one plane" is described as quoted. Separate
items are given for each different radius and the radius is stated in item
descriptions. Width ranges and units of measurement are set out in Table 8.04.

Formwork to curved surfaces other than that to one radius in one plane is
classed as "other curved". Item descriptions state the shape of such surfaces
and state the radius or radii, as the case may be, in accordance with Note G6
of the CESMM. Separate items are given for each different shape. Formwork of
this classification is measured by area and is given in m2, irrespective of width.

Where formwork is to curved surfaces of a component which is of constant
cross-section, it is considered preferable to measure it by length in accordance
with the alternative method given in Note G4 of the CESMM, rather than to
measure the surfaces as outlined above. See note against "For concrete components
of constant cross-section" in Table 8.04. See also subsequent Commentary.

Formwork to Voids

The classification "formwork for voids" is applicable to the formwork required
to form holes, pockets and the like within the limits of area or diameter stated
in Note G7 of the CESMM. Voids are described as either large or small voids as
defined in Note G7 of the CESMM. The formwork to them is given as numbered
items. Descriptions will usually state the types of voids, e.g. pockets, sumps,
etc. Separate items are given for each type. Further itemisation with particular
identifying descriptions may be given, where in the interest of clarity or of
distinguishing meaningful cost differences it is considered necessary to do so.
Measurement conventions and other details are given against the descriptive
feature in Table 8.04.

Formwork for Concrete Components of Constant Cross-Section

The classification "formwork for concrete components of constant cross-section"
may be used instead of the other formwork classificaions where the formwork for
a component or part of a component is of constant girth and profile when viewed
in cross-section. The alternative it provides is instead of measuring items
for the formwork for each of the surfaces of the component the combined surfaces
may be given as one item measured by length subject to Note G4 of the CESMM.
The Note requires particular information to be included in the item description.
See notes against the classification in Table 8.04.

Items of formwork classified and given in accordance with the alternative,
convey more clearly the nature of the formwork and the extent of repetition
than would be apparent from a series of items measured and given in accordance
with other formwork classifications. In practice when taking measurements of
formwork for concrete components of constant cross-section for the preparation of
the Bill, awareness of the advantages of measuring the formwork by length in
accordance with the alternative leads to it being used wherever practical as the
preferred method of measurement.

It is of advantage to use the alternative and measure by length as one
item the formwork for rebates, grooves and fillets, where otherwise, in
accordance with Note G5 of the CESMM, the formwork to the several surface
planes would be measured as a series of plane formwork items. The alternative
has broad application and may be used not only for the formwork for components
such as columns, beams and the like, but also for formwork for walls and that
for features of all kinds, provided the formwork is of constant girth and profile

Table 8.05   Reinforcement

Generally - Mass of steel reinforcement shall be taken as 0.785 kg/m per 100 mm2 of cross-section (7.85 t/m3). Mass of other reinforcing material shall be specified (Note G10)

1st Division	Reinforcement		
2nd Division			**3rd Division**
Mild steel bars to BS 4449 t High yield steel bars to BS 4449 t Stainless steel bars of stated quality t Reinforcing bars of other stated material t	Include mass of steel supports to top reinforcement (Note G10) Classify bars which are not circular in cross-section as nearest circular diameter listed which is nearest in cross-sectional area (Note G11) State length to next higher multiple of 3 m where bars exceed 12 m in length before bending (Note G12)		Classify according to diameter of bars. Group together in one item bars of 25 mm diameter and greater
High yield steel fabric to BS 4483 m2	State type number of BS (Note G13)	Measure nett with no allowance for laps (Note G13)	Classify according to range in kg/m2 State actual mass per m2 where it exceeds 8 kg/m2 and where only one mass per m2 in one item
Fabric of other stated material m2	State the material, sizes and its nominal mass per m2 (Note G13)		

COMMENTARY

Reinforcement (Refer to Table 8.05)

Item descriptions for reinforcement state the materials to be used and where there are alternative qualities, their quality. For steel reinforcement to B.S., it is sufficient to state the B.S. reference of the reinforcement.

In the normal course, items in accordance with the CESMM for reinforcement for a structure given in the particular Part of the Bill appropriate to the structure, is considered sufficient to identify the reinforcement. This is not to say that situations do not arise where it is reasonable to provide additional itemisation to distinguish cost differences. For example helical reinforcement and reinforcement to special profile to suit the shape of a component would be so described and be given separately from reinforcement which involved only normal bonding and hooking.

Bar Reinforcement

Bar reinforcement is measured by mass. Quantities give the calculated mass of the bars and that of any metal supports to top reinforcement. They exclude the mass of other supporting and tying reinforcement. For calculating mass, steel is taken as 7.85 t/m3, that of other reinforcement is taken as stated in the Contract.

COMMENTARY

Reinforcement (cont.

Bar Reinforcement (cont.

Separate items are given for each different diameter of bar less than 25 mm. Bars 25 mm diameter and greater are grouped in one item. Item descriptions for bars exceeding 12 m long, before bending, state the length of the next higher multiple of 3 m. The quantities of bars are given in tonnes, usually rounded off to the nearest tonne. Where it is considered unsatisfactory to round off quantities, fractional quantities may be used. See Paragraph 5.18 of the CESMM.

Fabric Reinforcement

Fabric reinforcement is measured by area in m2. The area of additional fabric in laps is not measured. The CESMM gives no direction as to the size of openings it is appropriate to deduct. Applying a strict interpretation, the area given is the area of the fabric, excluding the additional area of laps, left in the finished work. A practical application is to make no deduction for an opening in the fabric which is of an area not greater than that of a large void as defined in Note G7 of the CESMM. In the absence of a note in Preamble indicating the adoption of the convention mentioned in the preceding sentence, the Contractor must presume that fabric has been measured in strict accord with the provision of the CESMM.

Note G13 of the CESMM requires that item descriptions for high yield steel fabric to B.S. 4483 state the type number accorded the particular fabric in the B.S. Those for other fabric reinforcement are required to state material, sizes and nominal mass per square metre. Compliance with the Note creates items which state or reflect actual mass per square metre in place of a Third Division range.

Specimen Item Descriptions for Reinforcement

The specimen descriptions envisage all reinforcement to be steel to BS. The sub-headings would state materials and quality if it was otherwise.

Because of its special nature, helical reinforcement is located and given separately from that requiring only the normal bending and hooking.

Bars over 12 m long before bending, are given to the next higher multiple of 3 m. The bars in items Code G526.1 and 2 are 12 – 15 and 15 – 18 metres long, respectively.

Each different type of fabric reinforcement of a particular nominal mass is given as a separate item in consequence of Note G13 of the CESMM. See Table 8.05

Number	Item description	Unit
	CONCRETE ANCILLARIES	
	Reinforcement	
	Mild steel bars to BS 4449, grade 250	
G512.1	Diameter 8 mm.	t
G512.2	Diameter 8 mm; helical binding, columns E6 – E34.	t
	High yield steel bars to BS 4449, grade 460/425	
G525	Diameter 16 mm.	t
G526.1	Diameter 20 mm, length 15 m.	t
G526.2	Diameter 20 mm, length 18 m.	t
	High yield steel fabric to BS 4483	
G553	Reference No. A252, nominal mass 3.95 kg/m2	m2

Table 8.06   Joints

1st Division - Joints

Generally - Separate items are not required for formwork to joints or for joining or cutting waterstops (Note G14)

2nd Division - Surfaces

The classification "Formed surface" applies to joints requiring temporary support of the whole surface area of the concrete during casting. Other joints are classed as "Open surface". (Note G15)

Open surface m2	plain		Measure width or depth between outer faces of concrete with no addition or deduction for face or internal details (Note G16)
Formed surface m2	with filler	State dimensions and nature of filler (Note G14)	

3rd Division - Classify according to average width or depth of each continuous joint (Note G16)

2nd Division - Internal and external details			3rd Division
Plastic or rubber waterstops m		State spacing, dimensions and nature of components (Note G14)	Classify in accordance with 3rd Division descriptive features
Metal waterstops m			
Sealed rebate or groove m			
Dowel assembly nr			

COMMENTARY

Joints (refer to Table 8.06)

Two categories of items are given for joints, those which relate to work to the surfaces of the joints and those which relate to any components required for the joints. The latter being classed as internal and external details.

Items for joint surfaces are measured where joints are expressly required to be made, with or without formwork and with or without filler material. Joint surfaces are measured by area in m2. The conventions used for measuring width for the purpose of calculating area and for classification purposes are given in Note G16 of the CESMM. The classifications "open surface joints" and "formed surface joints" are defined in Note G15 of the CESMM. Any filler material required to the joints is included in the description of the joints. The dimensions and the nature of the filler is stated in the descriptions. See Table 8.06 for notes on the CESMM Notes mentioned in this paragraph.

Where formwork is necessary to form the joint surface it is measured in accordance with Note G2 of the CESMM. The provisions of Note G14 of the CESMM do not preclude the measurement of formwork to joints, the Note merely makes it unnecessary to state that such formwork is to joints.

Internal details include waterstops and dowel assemblies incorporated in joints. External details include sealing rebates and grooves in the faces of

COMMENTARY

Joints (cont.

the joints. Item descriptions for internal and external details state or
otherwise identify the nature of the components, including materials and
type, and the dimensions and spacing of the components. Waterstops and
sealed rebates and grooves are measured in linear metres. Dowel assemblies
are enumerated.   Unless otherwise required, items for internal and external
details are worded to include the provision of the components.

Example FE.1., included subsequently in this Chapter, provides a
measured example. In the Example the application of the provisions of the
CESMM to the measurement of joints is illustrated.

Table 8.07   Post-tensioned Prestressing

Generally   Separate items are not required for ducts, anchorages, grouting or
other components or tasks ancillary to prestressing (Note G19).

1st Division   Post-tensioned prestressing

2nd Division			3rd Division
Horizontal internal tendons in in situ concrete	nr	Measure by number of tendons (Note G18)	Measure developed lengths of tendons between outer faces of anchorages (Note G19)
Inclined and vertical internal tendons in in situ concrete	nr	Identify the concrete component to be stressed (Note G17)	
Horizontal internal tendons in precast concrete	nr	State the composition of tendon and particulars of anchorage (Note G17)	Classify according to appropriate length range in classification table, or state actual developed length where only one length in one item
Inclined and vertical internal tendons in precast concrete	nr	Classify profiled tendons in horizontal components as "horizontal tendons" (Note G19)	
External jacking operations	nr	Measure by number of external jacking operations (Note G18)	
		Identify the concrete component to be stressed (Note G17)	

COMMENTARY

Post-tensioned Prestressing (Refer to Table 8.07)

Prestressing is measured by numbering the tendons, where tendons are used, and
by numbering the external jacking points where stress is induced by jacking only.
Elaborate descriptions are not given. Item descriptions make reference to the
Specification where the prestressing system and the stressing components are
described and to the Drawings showing the concrete components and other details.

An example follows of a Bill description for the post-tensioned prestressing
of in situ concrete beams using tendons.

COMMENTARY

Post-tensioned Prestressing (cont.

Specimen Item Descriptions for Post-tensioned Prestressing

The item description uses the 1st and 2nd Division standard features and gives:-

the developed length of the tendons (Note G19 of CESMM), expressed as within a 3rd Division range. Had there been only one length in one item actual developed length would have been given

the composition of the tendons (Note G17 of CESMM)

particulars of anchorages, by Specification reference (Note G17 of CESMM)

identity of components to be stressed (Note G17 of CESMM)

Number	Item description	Unit
	CONCRETE ANCILLARIES Post-tensioned prestressing, as Specification clauses 18.1 to 18.10	
G717	Horizontal internal tendons in in situ concrete, length 25 - 30 m; of 7/12 mm diameter strands with anchorages as Specification clause 18.05; main beams M1 to M5 Drawing No. 197/22.	nr

Table 8.08   Concrete Accessories

1st Division	Concrete accessories	
2nd Division		3rd Division
Finishing of top surfaces m2 Finishing of formed surfaces m2	Measure only when separate finishing treatment required (Note G20) No deductions for openings less than 0.5 m2 in area (Note G20)	State required surface finish. State materials and thickness of applied finish (Note G20)
Inserts	All components cast or grouted into in situ concrete except reinforcement, structural metal-work, prestressing and jointing materials are classed as inserts (Note G21)	Linear inserts m Other inserts nr

COMMENTARY

Concrete Accessories (Refer to Table 8.08)

Finishing of Surfaces

Items for finishing surfaces of concrete are measured where a worked finish (as opposed to that produced by casting the concrete against formwork designed to give the desired finish) or where an applied finish is required. Finishing top surfaces is given separately from that to formed surfaces. See Table 8.08 for descriptive features, rules and units of measurement and additional description to be stated. Top surfaces which are not given separate treatment are not measured for finishing of top surfaces.

COMMENTARY

Concrete Accessories (cont.

Inserts

For work which is classed as "Inserts" see Table 8.08

Specimen Item Descriptions for Finishing of Surfaces and Inserts Number		Item description	Unit
Item descriptions for applied finishing state the material and the thickness. This is illustrated in the specimen description Code 813. In cases where the applied finishing is of varying thickness, average thickness is usually appropriate.		CONCRETE ANCILLARIES	
		Concrete Accessories	
		Finishing of top surfaces	
	G812	Steel trowel finish; tops of walls	m2
	G813	Granolithic finish, thickness 50 mm as Specification clause 16.07 including surface hardener as Specification clause 16.11; paved areas.	m2
There are three possibilities in the relation to work to be included in items for inserts. Items may be required to cover:-		Finishing of formed surfaces	
(i) supplying, fixing and casting in something not elsewhere given and required to be included in the item for the insert, or	G821	Aggregate exposure using retarder, as Specification clause 16.17; surfaces of plinths, flood walls as Drawing 107/17.	m2
(ii) fixing and casting in only something elsewhere given as supply only, or		Inserts including the components stated and casting in	
(iii) casting in only something elsewhere given as supply and fix.	G832.1	P.v.c. pipes, diameter 75 mm, length 300 mm, in reinforced walls; weep holes.	nr
Specimen item descriptions cover each of the three possibilities. See adjoining items Codes from G832.1 onwards.		Inserts and fixing and casting in only (supply of components elsewhere given)	
It will be noted from the specimen item descriptions for inserts that they identify the components to be cast in. They indicate the extent of the work in the items. The mention of reinforced walls, etc., infers there will be work to the formwork. More complicated work to the formwork, if it arises in connection with inserts, would be specifically stated in the description. Ad-hoc dimension ranges are used in items Codes G832.3 and 4 to avoid numerous items, of little cost difference, which using actual dimensions would create.	G831	Sliding door tracks in ground slabs (4 nr doors).	m
	G832.2	Set of four anchor bolts, including supplying and fixing sleeves and grouting sleeves and beneath stanchion baseplate, as Drawing No. 107/33.	nr
		Inserts and casting in only (supply and fixing of components elsewhere given)	
	G832.3	Pipes, nominal bore not exceeding 200 mm, through reinforced walls, thickness 300 - 600 mm.	nr
	G832.4	Pipes, nominal bore 200 - 300 mm, through reinforced suspended slabs, thickness 150 - 300 mm.	nr

Class H covers the manufacture, erection, joining and fixing of precast concrete units. The majority of the descriptive features in the Class relate to structural units. Among the specific exclusions from the Class are pipework, manholes, piles, paving units, kerbs, traffic sign supports, tunnel linings, blockwork and fencing of precast concrete. This excluded work is measured as provided in other Classes of the CESMM. For work included in the Class and for exclusions from the Class refer to the "Includes" and "Excludes" at the head of the Class H classification table in the CESMM. Cross references are also made to work included in Class H in the "Excludes" at the head of the classification tables in CESMM Classes G, M, N and W. The cross references in Classes G and M confirm that the items in Class H are to include moulds, formwork, reinforcement and any pre-tensioned prestressing required for the units. Those in Classes N and W make it clear that any metal inserts and waterproof joints identified with the units are to form part of the items for the precast units.

Table 8.09 Precast Concrete

Generally — State specification of concrete (Note H1) State position in the Works of units (Note H1) Units are deemed to be reinforced but not prestressed (Note H1) Give particulars of prestressing for prestressed units (Note H1)

Components cast in other than their final position are generally classed as precast (Note H4) Class as in situ concrete the site precasting of units where work is characteristic of in situ but requires movement into final position after casting and give movement operations as items in Temporary Works Class A (Note H4)

1st Division — Units with different dimensions shall be given different mark or type numbers (Note H5)			2nd Division	3rd Division
Beams	nr	State cross-sectional type and principal dimensions of cross-section (Note H2) State mark or type number (Note H3)	State length	State mass of each unit, or where length is to be determined by the Contractor state mass per m (Note H6)
Prestressed pre-tensioned beams	nr			
Prestressed post-tensioned beams	nr			
Columns	nr			
Slabs	nr	State average thickness (Note H2) State mark or type number (Note H3)	State area	
Segmental units	nr	State cross-section type and principal dimensions of cross-section (Note H2) State mark or type number (Note H3)		
Box units for subways, culverts and ducts	m			
The measured length for box units shall be the total length of identical units (Note H6)				

Table 8.09 Precast Concrete (cont.

1st Division		2nd Division	3rd Division
Copings, sills and wier blocks m The measured length shall be the total length of identical units (Note H6) State where wier blocks are laid to precise levels (Note H7)	State cross-sectional type and principal dimensions of cross-section (Note H2) State mark or type number (Note H3)	State cross-sectional area	State mass of each unit, or where length is to be determined by the Contractor state mass per m (Note H6)

COMMENTARY

Precast Concrete (Refer to Table 8.09)

Itemisation in accordance with Class H of the CESMM provides items for each
different size of each different type of precast concrete unit. The units
included in any one item must be identical. The quantities for box units for
subways, culverts and ducts and also those for copings, sills and wier blocks
are given in linear metres. Those for other classifications give the number of
units. Table 8.09 sets out the First Division descriptive features and notes
the units of measurement and the details required to be stated in item descriptions. In consequence of the provisions of Notes H3 and H5 of the CESMM, actual
length, area, cross-sectional area and mass, as the case may be, of the units
are given in item descriptions in place of the ranges given at Second and Third
Division level in the classification table of Class H.

Specimen Item Descriptions for Precast Concrete	Number	Item Description	Unit
The specimen item descriptions for beams are given to comply with Class H of CESMM. They state:-		PRECAST CONCRETE Design mix for ordinary structural concrete, grade 40, as Specification clause "X"	
the specification of the concrete, the position in the Works of each unit, the identity of the units, cross-sectional type,	H113	Moorlands Bridge Deck Secondary beams, inverted tee, 180 x 420 mm, length 4.00 m, mass 0.55 t, mark SB2.	nr
principal cross-sectional dimensions, length, mass of each unit, and particulars of prestressing (for prestressed units)	H368	Prestressed post-tensioned main beams, I section, 800 x 1380 mm, length 22.50 m, mass 30t, mark B2, pre-stressing as Drawing No. "Y" and Specifica-tion clause "W".	nr

MEASURED EXAMPLES

Three measured examples follow:

Mass Concrete Wall - Example FE.1.

The Example sets out the dimensions taken for the simple concrete wall shown on Drawing F/D/1. It illustrates the application of CESMM Classes F and G to the measurement of straightforward mass concrete and formwork. The Example is considered sufficiently uncomplicated to enable the dimensions to be easily followed without the aid of commentary. The Drawing has been somewhat elaborated to show joints, expressly required, in the concrete. The joints have been measured conforming to the rules and using the classifications given in the CESMM Class G. Related earthworks have been omitted from the Example.

Reinforced Concrete Structure - Example FE.2.

In Example FE.2 the dimensions are set out in the centre of the page to allow space for diagrams on one side and commentary on the other. For the purpose of the Example, the inclusion of the diagrams alongside the taking off which relates to them is considered preferable to providing an arrangement drawing with details on a separate page. It serves also to illustrate diagramatically the methodical approach to the measurement of the structure whereby it is broken down into its several elements and each is taken off and completed separately.

In the Example the lengths of the beams and columns have either been timesed or brought to a total in waste. On large structures it is more convenient to prepare schedules and collect together the lengths of components of similar classification and cross-sectional dimensions. One total for each is then transferred to the dimension sheets for multiplication by its cross-sectional dimensions.

Circular Mass Concrete Reservoir - Example FE.3

The Example FE.3 gives the complete taking off for the Circular Concrete Reservoir shown on Drawing F/D/3, with the exception of that for the joints in the concrete and that for associated pipework. An example of measuring joints in concrete is given in Example FE.1. Measuring pipework is illustrated in Chapter 9.

Before commencing taking off, the methodical approach is to decide how best the work can first be broken down into sections or elements for the purpose of measurement. When this has been done the taking off is concentrated on one section or element until complete. Each is dealt with in a similar manner until the whole of the taking off is complete. The elemental break down used for the Example and the order of taking off is as follows:-

1. Bases and footings   4. Cross wall   7. Bank and seeding

2. Blinding and formation   5. General earthworks

3. Perimeter wall   6. Perimeter footpath

The choice of commencing with the taking of dimensions for the concrete work rather than commencing with the excavation and working through more or less in the order of construction is a personal one. The reason is that when taking dimensions from drawings it is considered convenient to take off and record the work which represents the dimensioned form of the work (in this case the concrete) and later to generate from the dimensions taken those which are appropriate to the excavation and other earthworks. Others may choose to adopt an order of take off different from that in the Example.

108

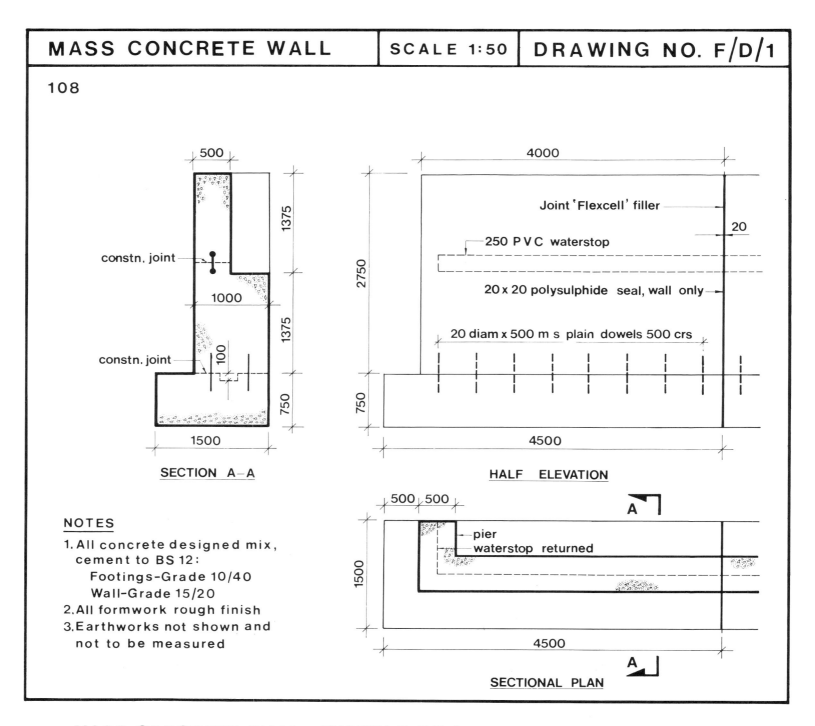

SECTION A–A

500

1375

constn. joint

1000

1375

constn. joint

100

750

1500

HALF ELEVATION

4000

Joint 'Flexcell' filler

20

2750

250 P V C waterstop

20 x 20 polysulphide seal, wall only

20 diam x 500 m s plain dowels 500 crs

750

4500

SECTIONAL PLAN

500 | 500

A

pier
waterstop returned

1500

4500

A

NOTES

1. All concrete designed mix, cement to BS 12:
   Footings-Grade 10/40
   Wall-Grade 15/20
2. All formwork rough finish
3. Earthworks not shown and not to be measured

MASS CONCRETE WALL – EXAMPLE FE.1

Note: All conc. desnd. mix ord. struc. conc. ct to BS.12

Footgs

2/	4.50
	1.50
	0.75

Provn. of conc. Grade 10., 40 agg. min. ct content 250 kg/m3

F114

&

Placg. of conc., mass footgs, thickn. e. 500.

F624

(1)

2/2/	4.50
	0.75
2/	1.50
	0.75

{ Fmw. rough fin., vert., width 0.4 – 1.22

(ends

G144

Wall lower section

2/	4.00
	1.00
	1.38

Provn. of conc., Grade 15, 20 agg. min. ct. content 290 kg/m3

F123

&

Placg. of conc. mass walls, thickn. e. 500

F644

(2)

Column (3)

Wall upper section

2/	4.00
	0.50
	1.38
2/	0.50
	0.50
	1.38

{ Provn. of conc., Grad. 15
20 agg., min. ct. content
290 kg/m3   F123.1

&   (piers

Placg. of conc., mass walls,
thickn 300-500
    F643

Fmw. to wall

2/2/	4.00
	2.75

Fmw. rough fin, vert., width
e. 1.22 m.   G145

2/	0.50
	1.38

Ddt. ditto   (pier back top

2/	1.00
	2.75
2/2/	0.50
	1.38

{ Fmw. rough fin., vert.,
width 0.4-1.22m (ends.
    G144
    (piers upper.

Har. const joint footg.

2/	4.00
	1.00

Joint surfs. Open surf.
plain, width 0.50-1m
    G612

2/2/	4.00
2/	0.50

Fmw. rough fin., vert.
width n.e. 0.1m (sides sinkg.
    G141

2/	8

Joints intl. & extl. details.
Dowel assembly plain, of 2 No.
20 diam × 500 mm M.S.
bars, 500 mm spacing
    G681.1

(3)

Column (4)

Hor. jt. at waterstop

2/	4.00
	0.50

Joint surf. - Open surf.
plain, width n.e. 0.5 m
    G611

	4000
½/500 =	250
	3750
	1000
½/500 =	250
	750

2/	3.75
2/	0.75

{ Joints intl & extl details -
Two bulb PVC waterstop,
width 250 mm
    G653.1

Vert. joint

0.50
1.38

Joint surfs. - Formed
surf. with 20 mm
"Flexcell" filler, width
n.e. 0.5 m   G641.1

1.00
1.38
1.50
0.75

{ Joint surfs - Formed
surf. with 20 mm
"Flexcell" filler, width
0.5 - 1m
    G642.1

0.50
1.38
1.00
1.38
1.50
0.75

{ Fmw. rough fin., vert.,
width 0.4 - 1.22 m
    G144

    (footg.

2.75

Joints intl & extl details -
Sealed surf. groove,
20 × 20 polysulphide
    G670.1

(4)

REINFORCED CONCRETE STRUCTURE – EXAMPLE F.E.2.

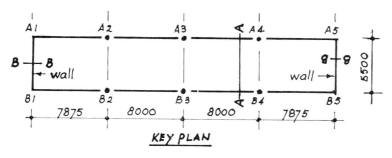

KEY PLAN

NOTES: All cols.   500 × 500
Edge beams. 1000 × 500
Walls.   250 thick
Slab.   350 thick

Key plan dimensions are to the
centre lines of cols and walls

Component	Dimensions	Commentary

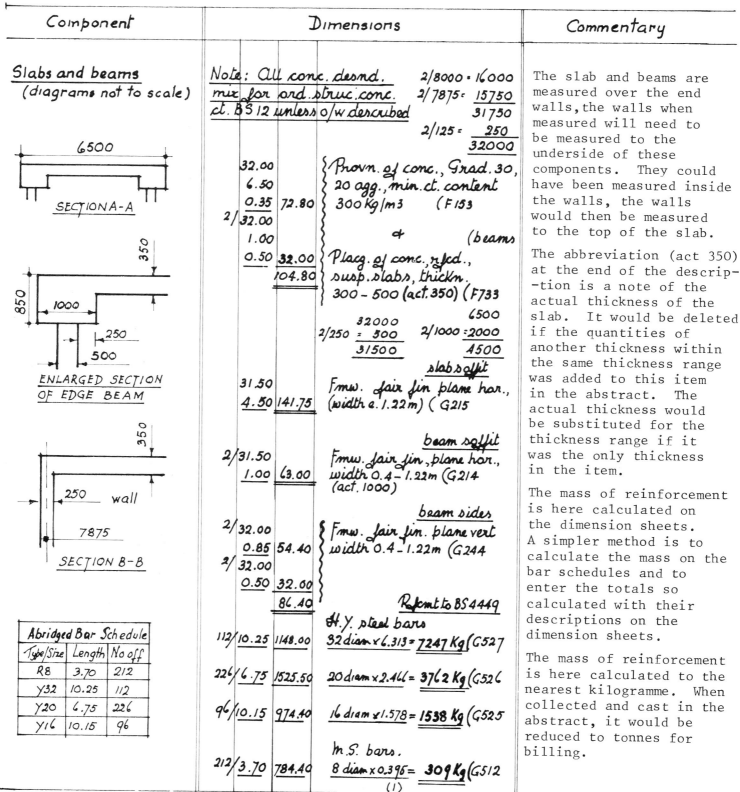

Slabs and beams
(diagrams not to scale)

SECTION A-A

ENLARGED SECTION
OF EDGE BEAM

SECTION B-B

Abridged Bar Schedule		
Type/Size	Length	No off
R8	3.70	212
Y32	10.25	112
Y20	6.75	226
Y16	10.15	96

Dimensions column:

Note: All conc. desnd. 2/8000 = 16000
mix for ord. struc. conc. 2/7875 = 15750
ct. BS 12 unless o/w described   31750
                             2/125 = 250
                                   32000

32.00
6.50
0.35 72.80
2/32.00
1.00
0.50 32.00
      104.80

Provn. of conc., Grad. 30,
20 agg., min. ct. content
300 Kg/m3   (F153

&   (beams

Placg. of conc., r fcd.,
susp. slabs, thickn.,
300 – 500 (act. 350) (F733

              6500
2/250 = 500   2/1000 = 2000
      31500   4500
             slab soffit

31.50
4.50 141.75
Fmw. fair fin plane hor.,
(width e. 1.22m) (G215

              beam soffit
2/31.50
1.00 63.00
Fmw. fair fin., plane hor.,
width 0.4 – 1.22m (G214
(act. 1000)

              beam sides
2/32.00
0.85 54.40
2/32.00
0.50 32.00
      86.40
Fmw. fair fin. plane vert
width 0.4 – 1.22m (G244

                    R fcmt to BS 4449
H.Y. steel bars
112/10.25 1148.00
32 diam × 6.313 = 7247 Kg (G527

226/6.75 1525.50
20 diam × 2.466 = 3762 Kg (G526

96/10.15 974.40
16 diam × 1.578 = 1538 Kg (G525

M.S. bars.
212/3.70 784.40
8 diam × 0.395 = 309 Kg (G512

(1)

Commentary column:

The slab and beams are measured over the end walls, the walls when measured will need to be measured to the underside of these components. They could have been measured inside the walls, the walls would then be measured to the top of the slab.

The abbreviation (act 350) at the end of the description is a note of the actual thickness of the slab. It would be deleted if the quantities of another thickness within the same thickness range was added to this item in the abstract. The actual thickness would be substituted for the thickness range if it was the only thickness in the item.

The mass of reinforcement is here calculated on the dimension sheets. A simpler method is to calculate the mass on the bar schedules and to enter the totals so calculated with their descriptions on the dimension sheets.

The mass of reinforcement is here calculated to the nearest kilogramme. When collected and cast in the abstract, it would be reduced to tonnes for billing.

Component	Dimensions	Commentary

Component column:

Columns 6 off
(diagrams not to scale)

— — — — —

beam

5000

ground level

1000

base

ELEVATION

500

500

PLAN

Abridged Bar Schedule

Type & Size	Length	No per col.	No off col/s
Y32	5725	6	6
R8	2000	17	6

Dimensions column:

<u>Cols</u>

2/3/ 0.50
0.50
5.00 7.50

Provn. of conc, Grad. 30, 20 agg.; min. ct content 300 kg/m3   (F 153

&

Placg. of conc., rfcd. cols. c.s.a. 0.1 – 0.25 m2; cols. A2 – A4 and B2 – B4 (act 0.25 m2)   (F 753

2/3/ 5.00 30.00

Fmw. fair fin., for conc. components of constant c.s. columns 500 × 500 mm, cols A2 – A4 and B2 – B4   (G 282

1000
75
925

2/3/ 0.93 5.58

Ddt. Fmw. as last item.   (G 282

&

Add Fmw rough fin, for conc. components of constant c.s. columns 500 × 500 mm, cols A2 – A4 and B2 – B4 below ground   (G 182

6/6/ 5.73
206.28 × 6.313 = 1302 kg (vert   (G 527

H. Y steel bars, diam. 32 mm

6/17/ 2.00
204.00 × 0.395 = 81 kg (links   (G 512

m. S. bars, diam. 8 mm

(2)

Commentary column:

It is important to state location of members in the Works where special characteristics may affect the method or rate of placing concrete. If this particular Example was billed as a separate Part of the Bill of Quantities, with a heading giving the title of the work, it is considered that members would be sufficiently located by the standard features which name the members, i.e. columns, suspended slabs, etc. Precise location has been added to item F753 merely as an example of how Note F2 of the CESMM would be implimented where it is wished to direct attention to a particular location which would not be clear without added description.

All columns being of constant cross-section the formwork is measured as a linear item according to the option provided in Note G4 of the CESMM. Location details are added to the description as required by the Note.

Formwork fair finish would not be required below ground. The "Ddt" and "Add" items substitute formwork rough finish for that part of the column which is below ground. Formwork fair finish is taken to the exposed surfaces of the columns and extended to 75 mm below ground level.

Component	Dimensions	Commentary

Component

Walls 2 off
(diagrams not to scale

6500

ELEVATION

250
5850

SECTION A-A

Abridged Bar Schedule		
Type/Size	Length	No. off
Y8	6.30	50
Y12	5.80	44

Dimensions

walls

```
2/ 6.50
   0.25
   5.00  16.25     { Prov^n of conc., Grad. 30,
2/ 4.50              20 agg ; min ct content
   0.25              300 Kg/m3        (F153
   0.50   1.13     {        &
         ─────     { Plac^g of conc., r^fcd.,
         17.38     { walls, thick^n 150-300 mm
                     (act 250)        (F742
```

```
2/ 6.50
   5.85  76.05    { Fmw. fair fin., plane vert.
2/ 6.50           { (width ≥ 1.22m)    (G245
   5.00  65.00    {
2/ 4.50           {     (inner face
   0.50   4.50    {
        ──────    {     ( "      "
        145.55
```

wall ends

```
2/2/ 0.25         Fmw fair fin., plane vert.
     5.00  5.00   width 0.2-0.4m (act 250)
                               (G243
```

rfcmt

H.Y steel bars

```
50/ 6.30  315.00   8 diam x 0.395 = 124 Kg (G522
44/ 5.80  255.20  12 diam x 0.888 = 227 Kg (G524
```

Commentary

The wall is measured to the underside of the slab and beams. See Commentary to the first column of dimensions in this Example. The formwork to the outer face of the wall is taken up to the top of this slab.

In the description for the formwork to the faces of the wall, "width exceeding 1.22 m" is noted merely as an aid to coding, which, in practice, may not be carried out at the same time as the dimensions are taken. The item description would read "Formwork fair finish vertical" in the Bill of Quantities. See Note G5 of the CESMM.

The whole of the formwork to the wall has been measured as fair finish. The formwork to the part of the wall below ground level (not shown) could be rough finish. If so, it should be given separately. Where the depth below ground is shallow, it may be considered reasonable to take the fair finish formwork down to the top of the foundation rather than give separately a narrow width of formwork rough finish.

Component	Dimensions	Commentary

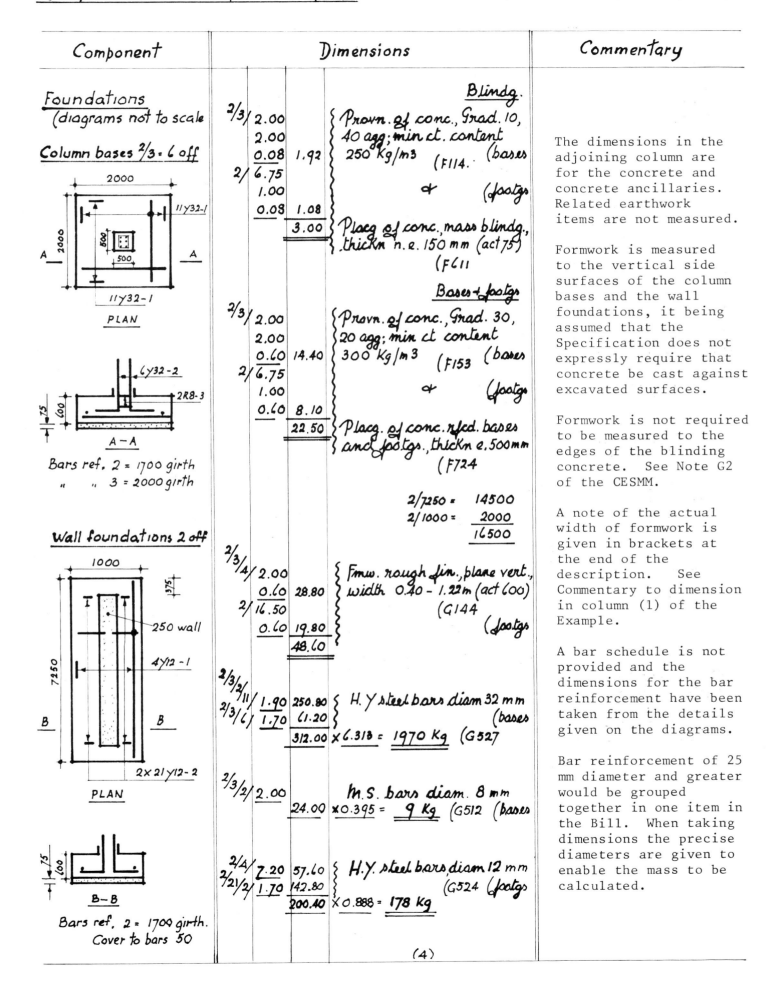

Foundations
(diagrams not to scale

Column bases 2/3 = 6 off

2000

2000

500

500

11 Y 32 - 1

11 Y 32 - 1

A   A

PLAN

6 Y 32 - 2

2 R 8 - 3

75   600

A — A

Bars ref. 2 = 1700 girth
"   " 3 = 2000 girth

Wall foundations 2 off

1000

375

250 wall

7250

4 Y 12 - 1

B   B

2 × 21 Y 12 - 2

PLAN

75   600

B — B

Bars ref. 2 = 1700 girth.
Cover to bars 50

Dimensions column:

Blindg.

2/3/ 2.00
     2.00
     0.08   1.92
2/   6.75
     1.00
     0.08   1.08
            3.00

Provn. of conc., Grad. 10,
40 agg; min. ct. content
250 kg/m3   (F114. (bases
                    & (footgs

Placg. of conc. mass blindg.,
thickn n.e. 150 mm (act 75)
                    (F611

Bases + footgs

2/3/ 2.00
     2.00
     0.60   14.40
2/   6.75
     1.00
     0.60   8.10
            22.50

Provn. of conc., Grad. 30,
20 agg; min ct content
300 kg/m3   (F153 (bases
                   & (footgs

Placg. of conc. rfcd. bases
and footgs., thickn e. 500 mm
                   (F724

2/7250 =   14500
2/1000 =   2000
           16500

2/3/4/ 2.00
       0.60   28.80
2/ 16.50
       0.60   19.80
              48.60

Fmw. rough fin., plane vert.,
width 0.40 - 1.22 m (act 600)
                    (G144
                    (footgs

2/3/2/11/ 1.90   250.80
2/3/6/   1.70   61.20
          312.00 × 6.313 = 1970 kg (G527

H.Y. steel bars diam 32 mm
                    (bases

2/3/2/ 2.00
       24.00 × 0.395 = 9 kg (G512 (bases

M.S. bars diam. 8 mm

2/4/ 7.20   57.60
2/21/2/ 1.70 142.80
        200.40 × 0.888 = 178 kg

H.Y. steel bars, diam 12 mm
                    (G524 (footgs

(4)

Commentary column:

The dimensions in the adjoining column are for the concrete and concrete ancillaries. Related earthwork items are not measured.

Formwork is measured to the vertical side surfaces of the column bases and the wall foundations, it being assumed that the Specification does not expressly require that concrete be cast against excavated surfaces.

Formwork is not required to be measured to the edges of the blinding concrete. See Note G2 of the CESMM.

A note of the actual width of formwork is given in brackets at the end of the description. See Commentary to dimension in column (1) of the Example.

A bar schedule is not provided and the dimensions for the bar reinforcement have been taken from the details given on the diagrams.

Bar reinforcement of 25 mm diameter and greater would be grouped together in one item in the Bill. When taking dimensions the precise diameters are given to enable the mass to be calculated.

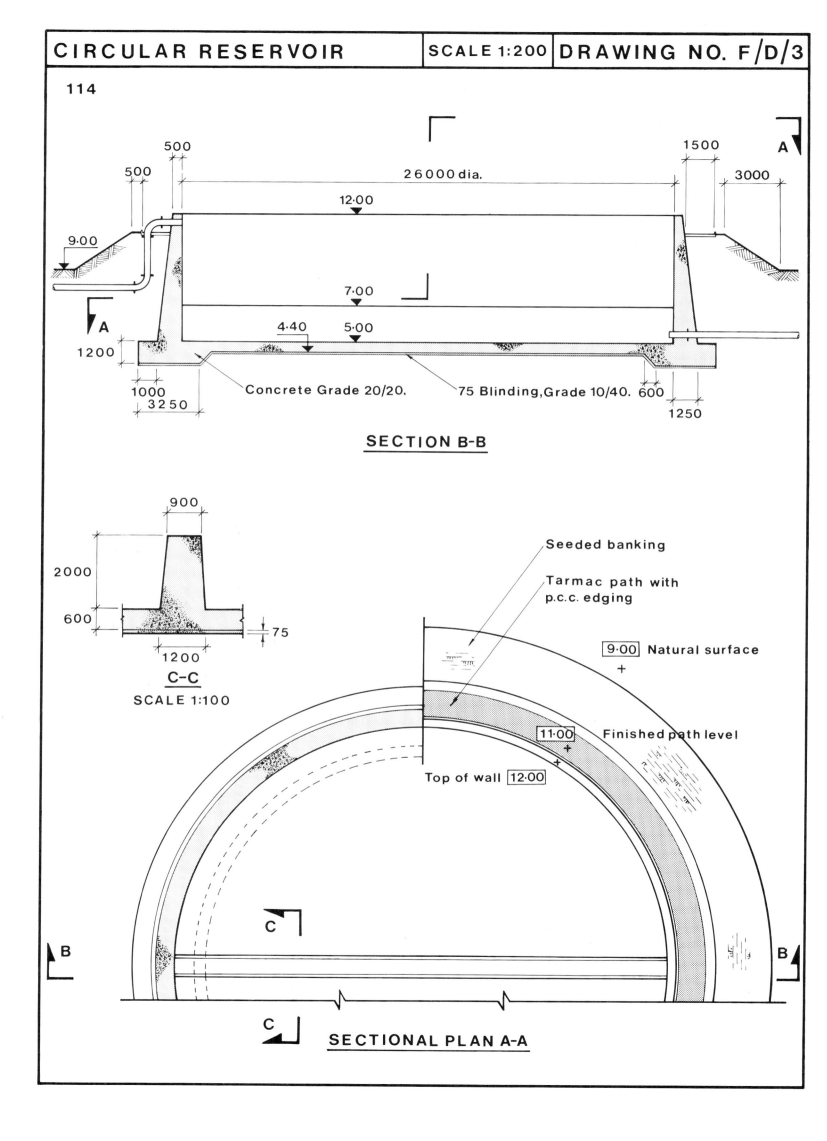

SECTION B-B

26000 dia.
500
500
1500
3000
12·00
9·00
7·00
5·00
4·40
1200
1000
3250
Concrete Grade 20/20.
75 Blinding, Grade 10/40.
600
1250

900
2000
600
1200
75

C–C
SCALE 1:100

Seeded banking

Tarmac path with p.c.c. edging

9·00 Natural surface

11·00

Finished path level

Top of wall 12·00

SECTIONAL PLAN A-A

CIRC. CONC. RESERVOIR   DRG. No. F/D/3

EXAMPLE FE.3

Note: All in situ conc. to be
desnd mix for ord. struc.
conc., ct. BS. 12, min. ct.
content 310 Kg/m3 unless o/w
described.

Bases & Footgs.

```
                    26000
        2/1250 = 2500
        2/1000 = 2000
diam. base  30500

                    26000
        2/1000 = 2000
                    24000
                    30500
                 2)54500
footgs.         27250
                    24000
        2/⅓/600 = 400
                    23600

        2/1000 = 2000
                    1250
                    3250

                 2)30500
radius          15250
```

COMMENTARY

All concrete other than blinding is one mix. The Note at the start of the taking off avoids having to repeat mix details in the descriptions.

The dimensions for the concrete base and footings are set down in the order noted at the right hand side of the description column. Preceding the dimensions the wastes calculate (1) the external diameter of the base (2) the mean diameter of the footings (3) the diameter at the centroid of the splayed haunch as noted in the following diagram.

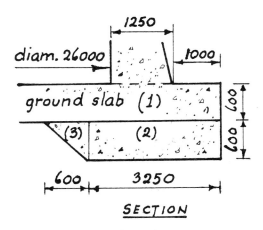

SECTION

The volume of the concrete thicknessing for the footing is added to that of the ground slab.

The thickness classification for placing the concrete is that of the ground slab.

```
22/  15.25
7/   15.25
     0.60
22/  27.25
7/   3.25
     0.60
½/22/  23.60
  7/   0.60
       0.60
```

{ Provn. of conc., Grad. 20,
20 agg.   (F133
                (g. slab.

            (footgs.

    &

            (splay.

{ Placg. of conc., mass
footgs & g. slab, thickn.
e. 500 mm (F624

CIRC. CONC. RESERVOIR   DRG. No. F/D/3

Bases & Footgs (cont.

22/7 /	13.00 13.00	Conc. access.. fin of top surfs.; spade fin. (G814
	26.00 1.20	Ddt. last item (G814 (cross wall.
22/7 /	30.50 1.20	Fmw. rough fin., curvd. one radius in one plane, width 0.4 – 1.22 m.; radius 15.25 m, perim of base (act. width 1.2m) (G154.1

Blindg & Formation

$$600 \times 1.414 = \underline{848}$$

22/7 /	15.25 15.25 0.08	Provn. of conc., prescbd. mix for ord. struc. conc., Grad. 10, ct BS12, 40 agg. (F214
22/7 /	23.40 0.85 0.08	(slope

+

Placg. of conc., mass blindg., thickn. n.e. 150 mm (act 75)   (F611

$$2/\tfrac{1}{2}/600 = \begin{array}{r} 24000 \\ \underline{600} \\ \underline{23400} \end{array}$$

22/7 /	23.40 0.60 0.08	Ddt both last items (hor. replaced (by slope.

(2)

The first item in the adjoining column is for the surface finish to the top of the base which is assumed to be spade finish. The second item deducts the surface finish for the area occupied by the base of the cross wall. If preferred the deduction could be made when dealing with the cross wall.

The formwork is for the side surfaces of the base, it being assumed that the Specification does not expressly require that the concrete shall be cast against the excavated surfaces. It is helpful to indicate the location of curved formwork in item descriptions. The width of the formwork is that of the thickness of the mass base at the edges. Formwork to the edges of the blinding is not measurable. Width classification is applicable to formwork curved to one radius.

The multiplier 1.414 is the Secant of 45 degrees and is used to calculate the slope length of the splayed inner edge of the thick-nessing to the perimeter of the base.

The first set of dimensions for blinding cover the plan area of the base, the second set covers the sloping blinding. The work is, therefore overmeasured to the extent of the flat area replaced by the slope. The Deduct item adjusts this. Distinction has not been made between placing hori-zontal blinding and placing that to slopes, it being considered that the cost difference is in-sufficiently significant to merit separate itemisation. See next column of Commentary.

CIRC. CONC. RESERVOIR DRG. NO F/D/3

Blindg & Formation (cont.

$$\begin{array}{r} 24000 \\ 22800 \\ \hline 2\overline{)46800} \\ \hline 23400 \end{array}$$

$\frac{22}{7}$ /	23.40 0.85	Fmw. rough fin., curvd. varying radius (conical) max. radius 12.00 m, min. radius 11.40 m; upper surfs of slopg. blindg. (G160.1

Formation

$\frac{22}{7}$ /	15.25 15.25	Excavn. ancills., prepn. of surfs. nat. matl. other than rock (E521.1

$\frac{22}{7}$ /	23.40 0.60	Ddt. last item

$\frac{22}{7}$ /	23.40 0.85	Excavn. ancills, prepn of surfs. nat matl. other than rock; curvd and slopg. (E521.2

Walls. perim
$$\begin{array}{r} 26000 \\ 2/500 = 1000 \\ \hline 2\overline{)27000} \\ \text{radius top } 13500 \\ \\ 26000 \\ 2/1250 = 2500 \\ \hline 2\overline{)28500} \\ \text{radius base } 14250 \end{array}$$

$14.25^2 = 203.06$
$13.50^2 = 182.25$
$14.25 \times 13.50 = 192.38$
$\underline{577.69}$

(3)

Notwithstanding that the last sentence of Note G2 of the CESMM provides that formwork to blinding concrete shall not be measured, based on the first sentence of Note G2 and the penultimate sentence of Note G1 of the CESMM, it is considered appropriate to measure formwork to the upper surface of the sloping blinding. This is given in the first item in the adjoining column.

An item for the preparation of curved and sloping surfaces is given separately from that for horizontal surfaces in keeping with Paragraph 5.10 of the CESMM.

The reservoir above the ground slab if imagined to be filled solid represents the frustum of a cone. The method chosen to measure the volume of the concrete in the perimeter walls is to first measure the volume of this frustum and deduct from it the cylindrical void. See sketch in the Commentary to the next column of dimensions. In the waste at the foot of the adjoining column, the top and bottom radii are first established. The formula for the frustum of a cone =

$$\frac{22}{7} \times \frac{1}{3} \text{ height } (R^2 + r^2 + Rr)$$

where R is the large radius and r the small. Calculations which represent the part of the formula in brackets are carried out in the waste column. The resultant is carried to the dimension column and the application of the remainder of the formula is carried out in the dimension and timesing columns. See first item in the next column of dimensions.

	CIRC. CONC. RESERVOIR DRG.No.F/D/3	COMMENTARY

Walls. perim (cont

$\frac{22}{7}/\frac{1}{3}/$	577.69	
	7.00	
	1.00	

Provn of conc., Grad.20 20 agg.   (F133

+

Placg. of conc., mass walls, thickn. e.500 mm   (F644

$\frac{22}{7}/$	13.00
	13.00
	7.00

Ddt. both last items (core

$\frac{22}{7}/$	26.00
	7.00

Fmw. fair fin., curvd. to one radius in one plane, radius 13.00 m, intl face of perim wall (width e 1.22)   (G255

```
        900
       1200
    2) 2100
       1050
```

$2/$	1.05
	2.00

Ddt. last item. (G255 (ends of cross wall

Exposed extl face of perim wall.

```
                    1250
                     500
                     750

750/7000 = 0.107 = 6° 10'

     batter = 10.71%

                    1000
below fin. level     150
                    1150
```

```
                          27000
2/1150 × 10.71% =           246
   max. diam            27246
   min. diam            27000
                     2) 54246
mean diam =            27123

1150 × 1.006 = 1157 (slope lgth
Note: 1.006 = Secant of 6° 10'
```

(4)

The first set of dimensions in the adjoining column represent the volume of the frustum of a cone (referred to in Commentary to preceding column of dimensions). The dimensions entered against the Deduct are those for the volume of the cylindrical void. See sketch below. The third dimension of 1.00 in the first set of dimensions is entered merely to indicate cubic dimensions.

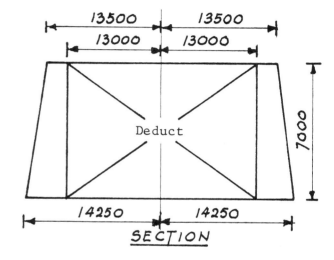

SECTION

Location of the curved formwork is given as additional description. Where the ends of the internal dividing walls abut the inside of the perimeter wall, the formwork to the latter is deducted. See Deduct item. Code G255.

For the purpose of measuring the formwork to the external face of the perimeter wall, the area which is above finished path level and that which is below needs to be separated. It being assumed that exposed external surfaces will have fair finish formwork. The waste at the foot of the adjoining column are the preliminary calculations to establish the dimensions of the formwork for the external exposed face of the perimeter wall. The calculations allow for the fair finish formwork extending to 150 mm below the finished path level. This is a practical allowance; it is not governed by a rule of CESMM.

CIRC. CONC.	RESERVOIR	DRG. NO. F/D/3	

<u>Walls, perim (cont.</u>

<u>Extl. face</u>

27246 ÷ 2 = <u>13623</u>

27000 ÷ 2 = <u>13500</u>

$\frac{22}{7}$ / 27.12
  1.16

Fmw. fair fin., curvd.
varying radius (conical)
max radius 13.62 m
min. radius 13.50 m,
extl face of perim wall
(G260

          7000
          1150
          <u>5850</u>

           27000
2/750 = 1500
          <u>28500</u>
          27246
2)<u>55746</u>
          27873

5850 x 1.006 = <u>5885</u>

$\frac{22}{7}$ / 27.88
  5.89

Fmw. rough fin., curvd.
varying radius (conical)
max radius 14.25 m
min. radius 13.62 m,
extl. face perim wall
below ground (G160.2

          27000
          26000
2)<u>53000</u>
          26500

<u>Top perim wall</u>

$\frac{22}{7}$ / 26.50
  0.50

Conc. access, fin of
top surfs., steel trow
fin.   (G812

(5)

COMMENTARY

The formwork to the outer face of the perimeter wall comes within the classification "other curved". It is described as provided in Note G6 of the CESMM and item descriptions state maximum and minimum radii. The first item is for the formwork above path level. The second item is for that below the level of the path. See sketch below.

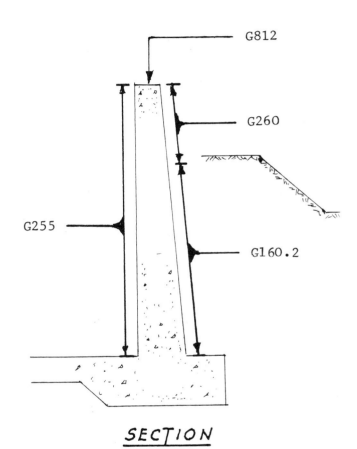

SECTION

An item is given for finishing the top surface of the wall on the assumption that it is specified to be steel trowelled finish. The dimensions represent the mean circumference of the top of the wall multiplied by the width.

CIRC. CONC. RESERVOIR — DRG. NO F/D/3

Walls. - Intl. cross wall

$$\begin{array}{r} 900 \\ 1200 \\ \hline 2)\,2100 \\ \hline 1050 \end{array}$$

26.00 1.05 2.00	Provn. of conc., Grad. 20 20 agg. (F133

&

Placg. of conc., mass walls, thickn. e 500 mm   (F644

$$\begin{array}{r} 1200 \\ 900 \\ \hline 2)\,300 \\ \hline 150 \end{array}$$

$$\frac{150}{2000} = 0.075 = 4°20'$$

$$2000 \times 1.003 = \underline{2006}$$

2/ 26.00 2.01	Fmw. fair fin., battd. (width e 1.22m) (G235

Top of cross wall

26.00 0.90	Conc. access., fin of top surfs., steel trow. fin. (G812

(6)

The section of the internal cross wall is as shown in the following diagram.

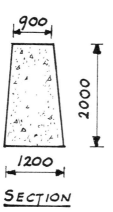

900

2000

1200

SECTION

The length of the internal wall is taken on its centre line. The slight reduction both as regards concrete and formwork due to the curved abutments is neglected.

The whole of the concrete in the wall exceeds 500 mm in thickness and is classified accordingly. Width for calculating volume is taken as the average width. See waste calculation.

The faces of the wall each have a batter of 150 mm in a height of 2.00 m. From these dimensions it is found that the angle of inclination is approximately 4° 20' to the vertical. The formwork being to a surface inclined at an angle within the range 0 - 10 degrees to the vertical is classified as "battered". The slope length of the formwork is calculated by multiplying the height of the wall by the Secant of 4° 20'.

It is assumed that the top of the wall is required to be steel trowelled finish and an item is measured for this.

CIRC. CONC. RESERVOIR. DRG. No. F/D/3

Earthworks for struc

Strip top soil.

$$
\begin{aligned}
&\quad 26000\\
2/500 &= 1000\\
2/1000 \times 10.71\% &= 214\\
2/2000 &= 4000\\
2/3000 &= 6000\\
2&\overline{)37214}\\
&\underline{18607}
\end{aligned}
$$

$\dfrac{22}{7}\bigg/$ 18.61
18.61
0.15

Gen. excavn., top soil for disposal, max. depth n.e. 0.25 m, Excavd. Surf. underside of top soil. (E421

Bulk dig.

$$
\begin{aligned}
&600\\
&\underline{75}\\
&\underline{675}\\[4pt]
&5000\\
&\underline{675}\\
\text{Formation.}\quad &\underline{4325}\\[4pt]
&9000\\
\text{Top soil.}\quad &\underline{150}\\
&8850\\
&4325\\
&\underline{4525}
\end{aligned}
$$

$\dfrac{22}{7}\bigg/$ 15.25
15.25
4.53

Gen. excavn. matl. for disposal, max depth 2 - 5 m, Comm Surf underside of top soil Excavd. Surf. 0.60 m above Fin. Surf. (E445

Note: Radius for preceding item from Column (1) this Example.

(7)

The first item measured for excavation assumes there is a specified requirement for top soil to be excavated and kept separate from the other excavated material. A separate stage is thereby created (See Note E7 of CESMM). The first item in the adjoining column of dimensions gives the excavation of the top soil as a separate stage of the general excavation and the description identifies the Excavated Surface and states the maximum depth of the stage. The diameter of the area of top soil is built up in waste and is divided by 2 to give radius. It is assumed soil is to be removed from beneath the perimeter banking. The extent is illustrated by dotted lines in the sketch given below.

General excavation below the top soil is given in the item under the heading "Bulk dig". It is measured over the spread of the foundation as provided in Note E8 of the CESMM. The depth (as calculated in waste) is measured to the level of the predominant formation. This excavation is indicated by cross hatching in the sketch below. As the Commencing Surface and the Excavated Surface for the work in the item are not the Original Surface and the Final Surface, respectively, each are identified in the item description.

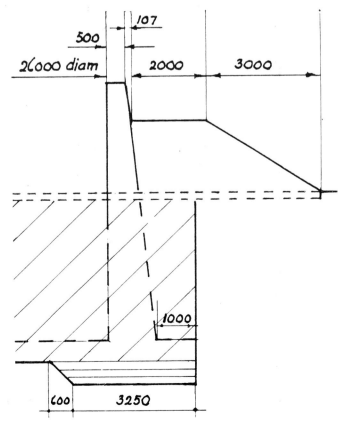

CIRC. CONC. RESERVOIR DRG. No F/D/3

<u>Perim. fdn.</u>

$75 \times 1.414 = 106$

2/31	23600	75
	62	31
	23538	24000

$2/\tfrac{1}{2}/31 = \dfrac{31}{23969}$

22/7	27.25	
	3.25	
	0.60	
½ /22/7	23.54	
	0.60	
	0.60	
22/7	23.97	
	0.03	
	0.60	

Excavn. of fdns. matl., for disposal, max. depth 0.5 – 1m; Comm. Surf. 0.60m above Final Surf. (E343

<u>Fill around perim walls</u>

2/½/1000 ·	30500
	1000
	29500

$3850 \times 10.71\% \cdot 412$

	29500
2/½/1000 ·	1000
2/¾/412 ·	275 1275
	28225

22/7	29.50	
	1.00	
	3.85	
½/22/7	28.23	
	0.41	
	3.85	

Fillg & compactn to strucs. sel excavd. matl. (E614

&

<u>Ddt.</u> Gen. excavn. matl. for disposal a.b.E445

&

<u>Add</u> Gen. excavn as last item but matl for re-use (E435

(8)

COMMENTARY

Excavation indicated by horizontal hatching in the sketch in the preceding column of Commentary is classified "Excavation of foundations". In accordance with Paragraph 5.21 of the CESMM, the description states the Commencing Surface. The first set of dimensions attached to the description is taken the same as that for the rectangular section of the concrete footing from Column (1) of this Example. The second set represents the triangular section of the excavation. The third set provides for the volume of the strip additional to the rectangular section as shown in section at (a) in the following sketch.

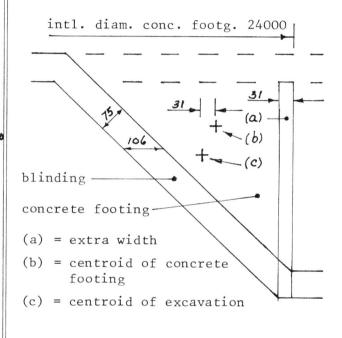

intl. diam. conc. footg. 24000

blinding

concrete footing

(a) = extra width

(b) = centroid of concrete footing

(c) = centroid of excavation

SECTION

The filling here measured fills the excavated void between the battered external face of the wall and the vertical excavated surface up to the underside of the top soil. For measurement convention, see Note E8 of the CESMM.

Following the filling item, general excavation previously taken as in material for disposal is re-classified as in material for re-use to the extent of the volume required for filling.

CIRC. CONC. RESERVOIR.   DRG. NO. F/D/3

Perim. Footpath.

```
                    27000
2/1000 x 10.71% =    214
                   27214
       2/1500 =     3000
                   30214
                   27214
                2) 57428
                   28714
```

22/7 / 28.71
     1.50

Li. duty pavements, bit. macadam, to BS.1621 Table 4, 6 mm. agg. depth 50mm. (R752

&

Li. duty pavements, granular base DoE Spec. clause 803, depth 100mm (R713

```
                    30214
           2/51 =    102
           2/75 =    150
                    30466
    1500            27214
      51         2) 57680
      75            28840
    1626
```

22/7 / 28.84
     1.63

Fillg ancills., prepn. of surfs., nat. matl. other than rock (E721

```
                30214
    2/½/51 =       51
                30265
```

22/7 / 30.27

Edgings, p.c.conc., to BS.340, Fig.11, 51 x 152mm, curvd. to radius 2.12 m, bedd. and haunchd. with conc. grad. 20, as detail Drg. No.8. (R441

(9)

The surfacing and base of the footpath is measured as provided in Class R of the CESMM. Class R is examined in Chapter 13.

The following diagram indicates the width dimensions of the footpath. The height of the wall above the footpath (1000) is multiplied by the rate of batter (10.71%) to calculate the dimension 107 mm.

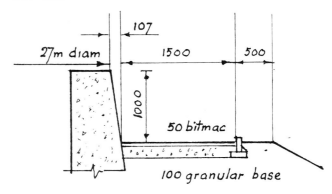

SECTION – PERIMETER PATH

The first set of waste calculations in the adjoining column of dimensions determines the diameter of the centre line of the footpath. The dimensions of the surfacing and the base are the mean girth of the footpath multiplied by the width. Measurements are taken at the top surface. See Note R3 of the CESMM.

Preparation of surfaces is measured beneath the base and the foundation to the edging.

The item description for edging makes reference to the detailed drawing reproduced below.

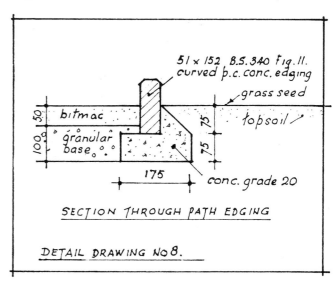

SECTION THROUGH PATH EDGING

DETAIL DRAWING No 8.

CIRC. CONC. RESERVOIR   DRG. No. F/D/3

COMMENTARY

Soil + seed banks to path

path level	11000
nat. g. level	9000
	2000

$$\frac{2000}{3000} = 0.666 = Tan$$
$$= \underline{33° 40'}$$

$3000 \times 1.202 = \underline{3605}$

```
            30214
2/500  =     1000
            31214
2/225  =      450
            30764

            31214
            37214
         2) 68428
            34214
```

Horiz. top

$\frac{22}{7}$ | 30.76
0.45

Fillg. + compactn. thickn. 150mm, excavd. soil (E631.1

Slopes

$\frac{22}{7}$ | 34.21
3.61

Fillg + compactn. thickn 150mm, excavd. soil, surfs. ov 10 degrees   (E631.2

Adjust excavn.

$\frac{22}{7}$ | 30.76
0.45
0.15

{ Ddt. Gen. excavn, soil for disposal a.b. E421   (top

$\frac{22}{7}$ | 34.21
3.61
0.15

+   (slopes

Add. Gen. excavn, all as last item, but soil for re-use (E411

Trim bulk fill for soil

$\frac{22}{7}$ | 34.21
3.61

Fillg. ancills, trimmg of slopes, nat. matl. (E711

(10)

The dimensions for the banking to the path are shown on the following diagram.

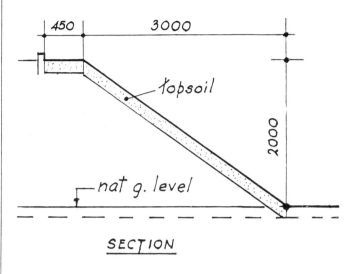

SECTION

Wastes first calculate height and then slope length. The angle at the toe of the bank is found either by calculator or by reference to Trig. Tables after the Tan. of the angle has been calculated by dividing height by the spread of the bank. The slope length is found by multiplying the spread by the Secant of the angle.

The first dimension in each set is the mean girth of the surface.

The uniform layer of topsoil is classified as "to stated thickness". That to surfaces inclined at an angle exceeding 10 degrees to the horizontal is given separately from that to horizontal surfaces.

Excavation of topsoil is given previously as for disposal and the volume required for the topsoil filling is re-classified by the "Ddt." and "Add" items.

Item E711 covers trimming the surface of the bulk filling to receive the uniform layer of topsoil.

<table>
<tr><td colspan="4">CIRC. CONC. RESERVOIR</td><td>DRG No F/D/3</td></tr>
</table>

Soil + seed banks (cont

Hor.

$\frac{22}{7}$ / 30.76
    0.45

Landscapg., grass seedg.,
surfs. n.e. 10 degrees.,
Spec. clause "X" (E831

Slope

$\frac{22}{7}$ / 34.21
    9.61

Landscapg., grass seedg
surfs ov. 10 degrees,
S.pec. clause "X" (E832

Bulk fillg. bankg.

150 × 1.202 = 180 − 150 = 30 mm
30 × 1.501 = 45 mm

diam. soil strip		37214
2/45 =		90
		37124
2/3000 =		6000
		31124

		27000
2/1150 × 10.71% =		246
		27246
2/2000 × 10.71%		428
		27674

		27674
		31124
	2)	58798
diam centroid (a)		29399

		31124
2/$\frac{1}{3}$/3000 =		2000
diam centroid (b)		33124

		27246
$\frac{2}{3}$/428 =		285
diam centroid (c)		27531

		2000
150 × 10.71% =		16
		1984
		45
		1939
2000 × 10.71% =		214
width (a)		1725

(11)

The dimensions for the grass seeding
are a repeat of those in the preceding column for the topsoil filling.

The first waste calculation under the
heading "Bulk fillg." uses trigonometry to find the difference between
the overall width of the bank and
that of the bulk filling.

1.202 = Secant of 33°40

1.501 = Cotan. of 33°40

The following diagram shows the
difference.

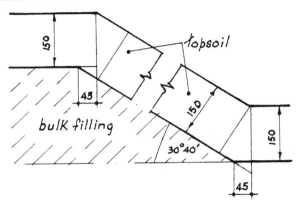

For the purpose of measurement the
cross-section of the bulk filling
is divided into a rectangle and two
triangles. See diagram below.

Further wastes are used to calculate
the minimum and maximum diameters of
the bulk filling. They are also used
to calculate the centroids, as
illustrated in the following diagram.

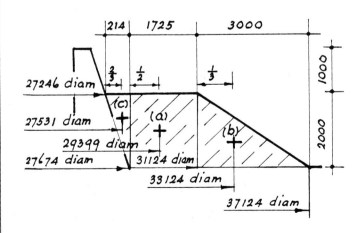

126

CIRC. CONC. RESERVOIR   DRG. No F/D/3

Bulk fillg. bankg. (cont.

$\frac{22}{7}$/ 29.40
    1.73
    2.00
$\frac{22}{7}$/$\frac{1}{2}$/ 33.12
    3.00
    2.00
$\frac{22}{7}$/$\frac{1}{2}$/ 27.53
    0.21
    2.00

Fillg. & compactn. to strucs. sel. excavd. matl. (E614   (a.

&   (b.

        (c.

Ddt. Gen. excavn. matl. for disposal a.b. E445

&

Add. Gen. excavn. matl. for re-use a.b. E435

(12)

COMMENTARY

The dimensions entered against the items in the adjoining column represent the circumference of the circles at the centroids multiplied by the cross-section for each of the sections (a), (b) and (c), respectively.

Notice that the filling to the banking is classified as "to structures".

A volume, equal to that of the filling, of excavation previously described as for disposal is re-classified as for re-use.

The abbreviation "a.b." followed by a code number, means as before described in the item with that code number.

9 Pipework—CESMM Classes: I-L

Each of the four Classes I - L of the CESMM are sub-divisions of the overall classification "Pipework". All four Classes are considered in this Chapter.

Class I covers the provision and jointing of pipes. Where the items describe the pipes as "in trenches" they include the laying of the pipes in the trenches and excavating and backfilling the pipe trenches. (See "Includes" at the head of the classification table Class I, CESMM).

The measurement of pipelines will usually start with the measurement of the pipes in accordance with Class I. The Class I items need to be complemented by items for the work covered by the other three Classes to the extent that such work is required in the pipelines. For example fittings and valves in the run of pipes are measured under Class J. Manholes and other chambers, gullies, crossings, the surface reinstatement of trenches and sundry ancillaries as listed in the table of classification, if required, are measured under Class K. Items measured under Classes I, J and K, which include excavation and backfilling, are deemed to include excavation in material other than rock or artificial hard material and backfilling with the material excavated. Any excavation in rock or artificial hard material and any backfilling with material other than that excavated, is given as items of extras to the previously included excavation and backfilling in accordance with Class L. Beds, haunches, surrounds and other supports required for the pipes together with any required wrapping of the pipes are measured under Class L. Items for pipe laying in headings and pipe laying by thrust boring, where expressly required, are given under Class L; these items are given in addition to the items for the provision and jointing of pipes "not in trenches" given under Class I. Excavation of headings is measured under Class T (See "Excludes" preceding the table of classification to Class E of the CESMM).

The principles of itemisation and measurement of "pipework" complemented by particular Notes, as set out in the CESMM apply to ducts and metal culverts. Class K provides classifications for the runs of ducts and metal culverts. Class K provides also features of classification for ditches and rubble drains and for the work additional to the pipes in trenches when French drains are required to be constructed.

For work included in Class I to L and for exclusions from the Classes, refer to the "Includes" and "Excludes" given at the head of the classification tables (for these Classes in the CESMM). References to work included in the Classes are also made in the "Excludes" for other Classes at the head of the classification tables in the CESMM as follows:-

References to work included in Classes I, J, K and L :-

Class E - Excavation for pipes and sewers, manholes, trenches and ditches.

Class M - Metalwork in pipework.

Class N - Pipework.

Class R - Drainage.

Class T - Pipe laying in headings, tunnels and shafts.

References to work included in Classes I and J:-

Class H - Precast concrete pipework.

References to work included in Classes L, J and K:-

Class W - Waterproofed joints.

References to work included in Class K:-

Class H - Precast concrete manholes, catchpits and
            gullies.

Class U - Brickwork to manholes and other brickwork
            incidental to pipework.

References to work included in Classes K and L:-

Class F - In-situ concrete for drainage and pipework.

Class G - Formwork and reinforcement ancillary to
            drainage and pipework.

Where a contract includes more than one type of pipeline, the work in each is grouped under a descriptive title. For example, Drainage, Water Mains, etc. Each may form a Part of the Bill of Quantities or a sub-division of a Part. Each sub-division may be further divided. For example, Drainage may be divided to group separately Sewers and Drains, French and Rubble Drains, etc. The appropriate items from all Classes being grouped in the Part or sub-division to which they relate.

PIPEWORK - PIPES - CESMM CLASS I

Item descriptions state the location or type of pipework so that pipe runs can be identified by reference to the Drawings. In addition, for work which is adjudged to comprise sections of differing characteristics, the Bill will group the work into sections giving separately the items for each section of work of particular characteristic. For example, items for work in open ground will be given separately from those for work in confined situations. Items for work in stable ground will be given separately from those for work in unstable ground, and so on.

COMMENTARY

Pipes (refer to Table 9.01)

Pipes are measured linearly, in metres, along their centre lines and include the lengths occupied by fittings and valves. Pipes which are built in to manholes and other chambers are measured from the inside surface of the manholes or chambers. For the measurement of pipes in backdrops to manholes, see subsequent Commentary on "Backdrops" under the heading "Manholes and Other Chambers" in Class K.

The first division descriptive features in Class I of the CESMM require to be amplified to the extent that item descriptions state the materials, the joint types, the lining requirements and any applicable BS reference and specified quality. The second division features are overridden by Note I3 which requires item descriptions to state the actual nominal bores of the pipes. Where pipe bores change at fittings, the usual convention is to measure the pipe of largest bore over the fitting. This is noted in Preamble so that the estimator knows the pipe cost he may offset against that of the fitting. Third division features distinguish pipes laid in trenches from those which are not.

Table 9.01 Pipes

Generally - Items are deemed to include all materials, unless otherwise stated (Note I1)

Separate items are not required for temporarily supporting sides of trenches, for keeping trenches free from water, for trimming trench bottoms or for disposal of surplus excavated material (Note I10)

1st Division			2nd Division
Clay pipes	m	State material, joint types, lining requirements,BS reference and quality (Note I3)	State nominal bore of pipe
Prestressed concrete pipes	m		
Other concrete pipes	m	Measure along centre line of pipes over fittings and valves, to inside surface of manholes and other chambers where built in (Note I5)	
Cast or spun iron pipes	m		
Steel pipes	m	State location or type of pipework so that runs can be identified by reference to the Drawings (Note I2)	
Plastic pipes	m		
Asbestos cement pipes	m	State where more than one pipe run in one trench (Note I4)	
Pitch fibre pipes	m	State where pipes laid in French or rubble drains (Note I4)	

3rd Division	
Not in trenches	Applies only where pipes expressly required not to be laid in trenches (Note I6)
	Applies to pipes suspended or supported above ground or other surface, to pipes in headings, tunnels and shafts, to pipes installed by thrust boring and pipes laid within volumes measured separately for excavation. Identify each such class (Note I6)
	Give additional items in Classes K and L for work in connection with pipes not in trenches (Note I6)
In trenches	State depth range in accordance with 3rd Division
	Measure depths from Commencing Surface to invert (Note I7)
	State trench depths exceeding 6 m to next higher multiple of 2 m (Note I8)
	Backfill is not measured, except as set out in Class K for French or rubble drains and except where backfilled with other than excavated material (Note I9)

COMMENTARY

Pipes (cont.

Pipes in Trenches

Items for pipes "in trenches" include the excavation of the trenches and for backfilling the trenches with the materials excavated. Separate items are not required for work as noted under "Generally" in Table 9.01

COMMENTARY

Pipes (cont.

Pipes in Trenches

The appropriate Third Division depth range within which the pipes are laid is measured from the Commencing Surface to the pipe invert. The Commencing Surface is identified in the item descriptions when it is not the Original Surface.

Separate items are given for each pipe when there is more than one pipe in one trench. The depth classification is that of the particular pipe. A statement of location combined with the phrase "in shared trench" in the item descriptions will usually identify pipes in these situations.

Items in addition to those in Class I are measurable under Class L when hard material is to be excavated, when backfilling is expressly required to be carried out in material other than that which is excavated, when surfaces require to be reinstated and when supports to sides of excavation are expressly required to be left in.

PIPEWORK — FITTINGS AND VALVES — CESMM CLASS J

Table 9.02 Fittings and Valves for Pipework

Generally	— Items are deemed to include all materials unless otherwise stated (Note J1)
	Separate items are not required for excavation, removal and disposal of material displaced by work in this Class (Note J2)
	Pipe fittings comprising backdrops to manholes shall be included in items for manholes in Class K (Note J7)

Fittings for pipework

1st Division		
Clay pipe fittings		State materials, joint types, lining requirements, BS reference and quality (Note J3)
Prestressed concrete pipe fittings		
Other concrete pipe fittings		
Cast or spun iron pipe fittings	State principal dimensions of fittings to cast iron pipework when nominal bore exceeds 300 mm (Note J5)	
Steel pipe fittings	State principal dimensions (Note J5)	
Plastic pipe fittings		
Asbestos cement pipe fittings		

2nd Division			3rd Division
Bends	nr	Measure vertical bends separately from horizontal bends in metal pipework (Note J5)	State nominal bore (Note J3)
Junctions and branches	nr		Classify pipe fittings comprising pipes of different bores according to the nominal bore of the largest pipe (Note J6)
Tapers	nr		
Double collars	nr		
Adaptors	nr		
Glands	nr		
Bellmouths	nr		

Table 9.02 Fittings and Valves for Pipework (cont.

Valves for Pipework

1st Division		2nd Division		3rd Division
Valves and penstocks	State materials and additional requirements such as joints, linings, drain cocks, extension spindles, BS reference and quality (Note J4)	Gate valves: hand operated	nr	State nominal bore (Note J4)
		power operated	nr	
		Non-return valves	nr	
		Butterfly valves:		
		hand operated	nr	
		power operated	nr	
		Air valves	nr	
		Penstocks	nr	

COMMENTARY

Fittings and Valves (refer to Table 9.02)

Fittings and valves are described and an item giving quantity by number is given in the Bill of Quantities for each type. The details to be stated in the item descriptions are given in Note J3. See Table 9.02. Actual nominal bores of valves and fittings are stated in the descriptions in place of the third division range. Fittings are classified according to their largest bore. Fittings comprising backdrops to manholes are considered to be part of the manhole and such fittings are not measured under Class J. The list of fittings and valves given in the classification table of Class J does not purport to be comprehensive. Fittings and valves other than those listed qualify to be measured.

PIPEWORK — MANHOLES AND PIPEWORK ANCILLARIES — CESMM   CLASS K

Separate items are not required for excavation for work in Class K, except as set out in Class L (See Notes 1 - 5 in Class L) (Note K4)

Table 9.03 Manholes, Catchpits and Gullies

1st Division		2nd Division		3rd Division
Manholes Catchpits	State type or mark numbers of manholes or other chambers details of which are given elsewhere in the Contract (Note K1)	Brick	nr	Measure depths from tops of covers to tops of base slabs (Note K3)
		*Brick with backdrop	nr	
		In situ concrete	nr	
	Identify different configurations of manholes and other chambers (Note K1)	*In situ concrete with backdrop	nr	State depth range as 3rd Division features, or State actual depth where only one in one item, or State actual depth where exceeding 6 m
		Precast concrete	nr	
	Separate items not required for different arrangements of inlets or for access shafts of different heights (Note K1)	*Precast concrete with backdrop	nr	
		*Note: Descriptive features marked thus * above, do not apply to catchpits		
	State types and loading duties of manhole and catchpit covers (Note K2)			
Gullies	State types and loading duties of covers (Note K2)	Classify stating 2nd Division descriptive features	nr	

COMMENTARY

Manholes and Other Chambers (refer to Table 9.03)

Manholes and other chambers are enumerated and are given by depth or depth range. See Table 9.03 for details. Separate items distinguish different forms of manholes and chambers. Separate items are not required for different arrangements of inlets or for access shafts of different heights. The loading duties of manhole covers are stated in item descriptions. Apart from extra for excavating any hard material and for any express requirement for backfilling with material other than that excavated (both of which are measured, if required, under Class L) the items represent the complete construction of the manholes and chambers as detailed in the Contract. Concrete or similar surrounds are detailed and made part of the manhole or chamber and are not classed as backfilling.

Depths of manholes and other chambers are measured from the tops of the covers to the tops of the base slabs. Where the benchings and bases of manholes are monolithic, it is necessary to state in Preamble the convention adopted in measuring the depths. Unless otherwise stated, it will be assumed that the Commencing Surface for the excavation is approximately that of the cover and that the Commencing Surface is the Original Surface.

Backdrops

Backdrop manholes are described as such. The pipes in the backdrops are measured linearly and are given separately from the main runs. They are described as "not in trenches". Additional description is given in the item stating the pipes are in "vertical backdrops within excavated volumes of manholes". Pipe fittings in the backdrops are deemed to be included in the backdrop manholes. (See CESMM Note J7).

In the items for backdrop manholes additional description or item coverage preamble is included making it clear that the prices for backdrop manholes are to include the extra depth of trench and the concrete surround for the pipes in the backdrops.

Table 9.04   French and Rubble Drains and Ditches

1st Division	2nd Division			3rd Division
French drains, rubble drains and ditches	Filling of French and rubble drains with granular material	m3	State nature of filling material (Note K5)	
	Filling of French and rubble drains with rubble	m3	Excavation and pipe laying shall be included in Class I (Note K5)	
	Trenches for unpiped rubble drains	m		Measure the cross-sectional area to the Excavated Surfaces (Note K7)
	Rectangular section ditches	unlined	m	
		lined	m State lining materials and dimensions (Note K7)	State cross-section area range as 3rd Division range, or State actual cross-section where only one in one item, or State actual cross-section where it exceeds 3 m2
	Vee section ditches	unlined	m	
		lined	m State lining materials and dimensions (Note K7)	

COMMENTARY

.French and Rubble Drains and Ditches (refer to Table 9.04)

Pipes and excavation for piped French and rubble drains are measured as provided in Class I of the CESMM. Trenches for unpiped rubble drains are measured under Class K, as indicated in Table 9.04. The filling of French and rubble drains is measured by volume and is given separately.

Lined and unlined ditches and also trenches for unpiped rubble drains are each classified, in the third division, according to cross-sectional area. The cross-section is measured from the Excavated Surfaces. Trenches and ditches which taper and extend into more than one cross-sectional area range are divided into separate items for each standard cross-sectional area range. Item descriptions for lined ditches state the lining material and its thickness and, in cases where the whole of the Excavated Surfaces are not lined, they define the extent of the lining.

Specimen Item Descriptions for French Drain

Number	Item description	Unit
I112	PIPEWORK - PIPES Clay land drain pipes as Specification clause 2.23, in French drains Nominal bore 150 mm, in trenches, depth not exceeding 1 m; Commencing Surface carriageway formation.	m
K410	PIPEWORK - MANHOLES AND PIPEWORK ANCILLARIES French drains Filling French drains with granular material, Specification clause 2.03	m3
L331	PIPEWORK - SUPPORTS AND PROTECTION, ANCILLARIES TO LAYING AND EXCAVATION Imported granular material as Specification clause 2.35 Beds depth 100 mm, nominal bore not exceeding 200 mm, Drawing No. PD/L.	m

FRENCH DRAINS - DRAWING PD/L.
(not to scale)

SECTION

COMMENTARY

Ducts and Metal Culverts (Refer to Table 9.05)

Ducts and metal culverts are measured in linear metres and are covered by descriptive features in Class K. See Table 9.05. The Notes in Classes I, J and L which refer to pipes and pipework apply to ducts and metal culverts. In item descriptions for non-circular metal ducts, the maximum nominal cross-sectional dimensions are substituted for the nominal internal diameter in the second division.

Table 9.05   Ducts and Metal Culverts

Generally - The Notes in Classes I, J and L*which refer to pipes and pipework
apply to ducts and metal culverts (Note K6)

1st Division	2nd Division	3rd Division
Ducts and metal culverts m	Use the 2nd Division descriptive features but apply the rules for pipes and pipework as given in the Notes for Classes I, J and L (Note K6) i.e. State materials, diameters, measure lengths and fittings, etc., all as provided for pipes and pipework. State maximum nominal cross-section dimensions for non circular metal ducts (Note K6) Separate items are not required for bends on cable ducts (Note K6)	Classify in accordance with the 3rd Division features as either:- Not in trenches, or In trenches, applying the same rules as for pipes and pipework

Table 9.06   Crossings, Reinstatement and Other Pipework Ancillaries

Generally - Items, etc., set out in Table 9.06 apply equally to pipes, pipework
metal ducts and culverts, except that dimensions for 3rd Division
classification of ducts and culverts shall be the maximum nominal
distance between the inside faces of the outer walls (Note K6)

1st Division	2nd Division			3rd Division
Crossings	River, stream or canal (width up to (10 m) nr	State width range as 2nd Division, or State actual width where only one in one item	Measure crossings of streams only when width exceeds 1 m (Note K9) Measure widths between banks along pipe centre lines (Note K9) Where linings are to be broken through and reinstated, state type of lining (Note K9)	State nominal bore range as 3rd Division features, or State actual nominal bore where only one in one item, or State actual nominal bore when it exceeds 2000 mm For ducts and culverts see Generally above
	River, stream or canal (width over 10 m) nr	State actual width		
	Hedge: Wall: Fence nr		Give each separately	

COMMENTARY

Crossings, Reinstatement and Other Pipework Ancillaries (Refer to Table 9.06)

The Commentary on crossings, reinstatement and other pipework ancillaries, which
follows, is applicable equally to pipes, pipework and ducts and metal culverts.

Crossings

A numbered item for crossings is given in addition to the pipes and other work
when a pipeline crosses a feature of the kind indicated in the second division
of "Crossings" in Table 9.06. Item descriptions for crossings of rivers, streams

* "I, J and L" should be substituted for "I, K and L" by an
   amendment in Preamble to amend the misprint in the first
   line of Note K6 of the CESMM.

COMMENTARY

Crossings, Reinstatement and Other Pipework Ancillaries (cont.

Crossings (cont.

and canals will usually state location in addition to the details required to be stated as noted in Table 9.06. The pipes in crossings are measured under Class I. The run of pipes in a crossing of a river, stream or canal would be given separately from the runs on the banks and item descriptions would give location and indicate that the work was affected by bodies of water. Other associated work such as supports, etc., when measured under the appropriate Class, which may be similarly affected would need to be so described in the item descriptions.

Table 9.06 Crossings, Reinstatement and Other Pipework ancillaries (cont.

1st Division	2nd Division			3rd Division
Reinstatement	Breaking up and temporary rein-statement of:- roads m footpaths m Breaking up and temporary and permanent rein-statement of:- roads m footpaths m	State type and maximum depth of surfaces (Note K8) Separate items not required for removal and rein-statement of kerbs (Note K8)	Measure lengths along centre lines of pipes and include lengths of manholes (Note K9)	State nominal bore range as 3rd Division features, or State actual nominal bore when only one in one item, or State actual nominal bore when it exceeds 2000 mm. For ducts and culverts, see "Generally" in first panel of Table 9.06
	Reinstatement: State type of surface m			
Other pipework ancillaries	Reinstatement of field drains nr			
	Marker posts nr	State size and type (Note K10)		
	Timber supports left in excavation m2 Metal supports left in excavation m2	Measure area of supported surface for which supports are expressly required to be left in (Note K11)		
	Connexions of pipes to existing manholes, chambers and pipes nr			

COMMENTARY

Crossings, Reinstatement and Other Pipework Ancillaries (cont.

Reinstatement

Reinstatement of surfaces excavated or broken through for pipelines is measured in linear metres. Reinstatement requirements are stated in the item descriptions. See Table 9.06. The measurement convention, providing that lengths are measured along the centre line of the pipes and include the lengths occupied by manholes (stated in the last sentence of Note K9), makes reference only to "breaking up and reinstatement of roads and footpaths". It is usual to measure the reinstatement of other surfaces (See 2nd Division Codes 5 - 8 of 1st Division Code K7) employing the same convention and to note this in Preamble.

COMMENTARY

Crossings, Reinstatement and Other Pipework Ancillaries (cont.

Connexions

Separate items are given for each type of connexion between pipework and existing work, each giving the number of connexions. Item descriptions give locational or other identifying information. They also state or identify associated work (such as the alteration of existing manholes and benchings or pipework and maintaining the flow of the existing during the making of the connexions) intended to be included in the items. Where connexions are made between new and existing pipes with fittings, such as junctions or saddles, the pipes, fittings, beds, etc., may be measured in the runs of the new pipework as provided in Classes I, J and L. The items for the connexions would then state that pipes, fittings, etc., are measured separately.

PIPEWORK – SUPPORTS AND PROTECTION, ANCILLARIES
TO LAYING AND EXCAVATION – CESMM CLASS L

Table 9.07   Class L, Generally

Generally – Items shall be given in this Class in addition to the items in Classes I and K as defined in Note L1.
Work associated with ducts and metal culverts shall be itemised, described and measured as set out for work associated with pipes, except that the dimension used for classification in the 3rd Division shall be as provided in Note L12.

Table 9.08   Extras to Excavation and Backfilling

1st Division – Extras to excavation and backfilling   m3

2nd Division

Excavation of rock (Define in Preamble) Excavation of mass concrete Excavation of reinforced concrete Excavation of other artificial hard material (State nature of material)			For trenches calculate volume as provided in Note L2 For manholes and other chambers calculate volume as provided in Note L3 Breaking up roads and pavings shall be included in Class K (Note L4)
Backfilling above Final Surface with concrete Backfilling above Final Surface with stated material other than concrete	Measure only if it is expressly required that material excavated shall not be used for backfilling (Note L5)	State grade of concrete if item would not otherwise identify State nature of material other than concrete.	
Excavation of natural material below Final Surface and backfilling with concrete Excavation of natural material below Final Surface and backfilling with stated material other than concrete	Measure only where expressly required (Note L5)		

COMMENTARY

Extras to Excavation and Backfilling (refer to Table 9.08)

Excavation in rock or other artificial hard material and backfilling with material other than that excavated are each classified, under Class L, as extras to the excavation and backfilling included in the items in Classes I and K. Items for the work give the quantities by volume in m3. (See Table 9.08). The volume of the work in trenches is determined using the average depth and length of the material removed or backfilled and the notional width of trench. The notional width of trench is that which is stated in the Contract. Where no such width is stated, the width is taken as the maximum distance between the inside faces of the outer pipe walls, plus 500 mm where this maximum distance does not exceed one metre, or plus 750 mm where this maximum distance exceeds one metre (See Note L2). The volume of extra for excavation of rock or artificial hard material in manholes and other chambers is calculated from the maximum horizontal cross-sectional area of the manhole or chamber and the average depth of the material removed. (See Note L3).

The measurement convention set out in Note L3 of the CESMM for determining the volume of backfilling for manholes and other chambers, in effect backfills the chamber and this is obviously a mistake. It is usual to include a statement in the Preamble of the Bill of Quantities which amends the CESMM in relation to this Note. An example of suggested amending preamble using hypothetical reference numbers follows:-

SECTION "B"

PREAMBLE

12.00   PIPEWORK - SUPPORTS PROTECTION, ANCILLARIES TO LAYING AND EXCAVATION

Amendments to CESMM, Class L

12.01   The undernoted amendments are deemed to have been made to Class L of the CESMM when preparing this Bill of Quantities.

12.02   Note L3, Page 35, deemed to be deleted.

12.03   Additional Notes as follows, deemed to be included in the Notes appended to Class L, on page 35.

12.04   Note 3(a) The volume of "extras to excavation and backfilling" for the excavation for "manholes" and other chambers shall be calculated by multiplying the average depth of the material removed by the maximum horizontal external cross-sectional area of the manhole or chamber.

12.05   Note 3(b) The volume of "extras to excavation and backfilling" for backfilling for "manholes" and other chambers shall be that of the compacted filled volume, assuming that the volume of excavation is deemed to have been determined as provided in Note 8 of Class E and provided that the backfilling is measured in accordance with Note 17 of Class E.

Table 9.09   Headings and Thrust Boring

1st Division – Headings and thrust boring			
2nd Division			**3rd Division**
Pipe laying in headings m	State type of packing (Note L6)	Measure only where expressly required (Note L6)	State nominal bore range as 3rd Division features: State actual nominal bore where only one in one item
Pipe laying by thrust boring m	Class thrust blocks as Temporary Works in Class A (Note L6)	Identify the run of pipe or pipes laid (Note L6)	

COMMENTARY

Headings and Thrust Boring (refer to Table 9.09)

Items for pipe laying in headings and for pipe laying by thrust boring are given in accordance with Class L (See Table 9.09), where such work is expressly required.   These items are given in addition to the Class I items for the pipes. In the Class I items, the pipes are classified "not in trenches" and are identified as being in headings or installed by thrust boring, as appropriate. Item descriptions for the Class L items of "pipe laying in headings" state the type of packing material and indicate that the items are to include packing with the stated packing material.

Excavation for headings and for pipes installed by thrust boring is treated as small tunnels measured in m3 in accordance with Class T.   (See Commentary under the heading "Excavation" in Chapter 15 and also the Commentary which follows).

Excavation of Headings

For the purpose of measurement, the Engineer must decide on the cross-section of a heading and prepare a drawing for it showing the payment lines proposed.   The volumes of excavation are calculated to the payment lines shown on the drawings. Items for excavated surfaces give the area in m2 at the payment lines.   (See Commentary on "Excavation" in Chapter 15 for the details to be given in item descriptions).

The Commentary on "Support and Stabilization" in Chapter 15 is applicable to timber or other support for headings.

Excavation for Pipes Installed by Thrust Boring

Preamble will note an amendment to Note T13 of the CESMM and it will be provided that the rates for excavation for pipes installed by thrust boring are to include for temporary ground support.

The cross-section used to calculate the volume of excavation is determined from the external diameter of the pipes.   The perimeter determined from this diameter is used to calculate the area of the excavated surfaces for the items for excavated surfaces.

COMMENTARY

Headings and Thrust Boring (cont.

Specimen Bill for Heading and Pipework

PART:6 STORMWATER SEWERS

Number	Item Description	Unit	Quantity	Rate	Amount	
					£	p
	Headings, Pipework in Heading under Yatton Road, Drawings Nos 29/7 - 9					
	PIPEWORK - PIPES					
	Vitrified clay pipes to BS.65/540:Part 1, with Mechanical Flexible Joints to BS.65/540:Part 2, not in trenches					
I121	Nominal Bore 300 mm, in headings.	m	23			
	PIPEWORK - SUPPORTS AND PROTECTION, ANCILLARIES TO LAYING AND EXCAVATION					
	Headings					
L212	Pipe laying in headings, nominal bore 300 mm, including packing with grade 10 concrete to completely fill headings.	m	23			
	TUNNELS					
	Excavation					
T121	Headings in boulder clay, cross-sectional dimensions 1075 x 1350 mm.	m3	34			
T180	Excavated surfaces in boulder clay and filling any voids caused by overbreak with grade 10 concrete.	m2	112			
	Support and stabilization;					
	Internal support					
	Timber supports, 100 x 100 mm sills and uprights, 100 x 125 mm heads					
T823	Supply.	m3	2			
T824	Erection.	m3	2			
T829	Steel dogs; for timber connections.	nr	112			
	Lagging					
T825	Timber poling boards, thickness 38 mm and packing behind with grade 10 concrete.	m2	95			
	To Part 6 Summary		Page total			

COMMENTARY

The CESMM makes no mention in Class T of metal components for connecting the timbers. It is considered more helpful to give the components as separate items rather than to include them in the items for the timber.

Where considered advisable items should be Billed for forward probing.

Table 9.10 Supports and Protection

1st Division	2nd Division		3rd Division
Generally – Where supports and protection are required to be measured linearly, measure length along pipe centre lines over fittings and valves but not including lengths occupied by manholes and chambers through which support or protection is not continued (Note L7)			
Beds m Haunches m Surrounds m	Sand Selected granular material Imported granular material Mass concrete Reinforced concrete	State materials used (Note L8) State depth of beds (Note L8) Separate items are not given for beds to haunched or surrounded pipes where same material is used for beds and haunches or beds and surrounds (Note L8)	State nominal bore range as 3rd Division features, or, State actual nominal bore where only one in one item, or, State actual nominal bore where it exceeds 1800 mm
Wrapping and lagging m	State material used (Note L9) Separate items are not required for wrapping and lagging fittings and valves (Note L9)		
Concrete stools and thrust blocks nr	State specification of concrete (Note L11) State volume range as 2nd Division features, or, State actual volume where only one in one item, or, State actual volume where it exceeds 8 m3		
Other isolated pipe supports nr	State principal dimensions. State materials used (Note L10) Measure height from ground or other supporting surface to invert of highest pipe where supported from below and to lowest pipe where pipes are supported from above. Where two pipes or more are carried by a support classify in 3rd Division by aggregate bore of pipes supported (Note L10)		

COMMENTARY

Supports and Protection (refer to Table 9.10)

Disposal of excavated material is not specifically included in the Class L items of the CESMM. Consequently, disposal requirements need to be stated, either by way of additional description or item coverage preamble, for the material resulting from the additional excavation for beds and for the material resulting from any excavation for stools, thrust blocks and isolated pipe supports.

Beds, Haunches and Surrounds

Beds, haunches and surrounds to pipework are measured and given in linear metres. Note L7 of the CESMM states the convention for determining lengths. (See "Generally" in Table 9.10.). Items for haunches and items for surrounds each include the related beds where the material used for them is the same as that used for the beds. Where material different from the bed is used for the haunches or surrounded section, these are each given separately from the beds and item descriptions state that the bed is measured separately.

Item descriptions state the materials used and the depths of the beds. They will identify the cross-sectional profiles of the beds haunches or surrounds (usually by drawing reference). Notwithstanding that the 3rd Division features provide for items to group nominal bores within ranges, separate items using the same nominal bore range, or stating a single nominal bore, may require to be given to distinguish cross-sectional profiles of differing cost characteristics.

COMMENTARY

Supports and Protection (cont.

Beds Haunches and Surrounds (cont.

Items include the additional excavation (beyond that included in the Class I items) to the extent that it is material other than rock or artificial hard material. Any rock or artificial hard material within the additional excavation which is to be removed is given by volume in accordance with the items in this Class for extras to excavation and backfilling.

Wrapping and Lagging

Wrapping and lagging pipes are measured and given in linear metres. The convention for determining lengths is given in Note L7 of the CESMM. Separate items are not given for wrapping or lagging fittings and valves. The 3rd Division features provide for the nominal bore of the pipes wrapped or lagged to be grouped in ranges. The assessed cost difference between work to pipes within a range may be considered such as to make it advisable to give items for each nominal bore in place of the nominal bore range.

Concrete Stools and Thrust Blocks

Item quantities give the number of concrete stools and thrust blocks. Item descriptions state the specification of the concrete and are required to state the volume range within which each stool or block falls and the nominal bore range of the related pipework. The items include excavation, formwork and reinforcement, if required, and item descriptions identify requirements in respect of these and any other requirements. The cost difference between the lower and higher volumes in the standard volume ranges could, in some cases, be substantial. Further itemisation is sometimes introduced to distinguish cost difference.

Enumerated items as here discussed are not appropriate for concrete stools of substantial construction. They would be dealt with as structures and measured and itemised in detail as provided in other Classes appropriate to the work.

Isolated Pipe Supports

Isolated pipe supports are enumerated. Item descriptions state the materials used and the principal dimensions of the supports. They state also the heights of the supports and also the nominal bore of the pipes supported within stated ranges. See Table 9.10. It is not appropriate to give substantial structure type pipe supports as enumerated items. They are measured and itemised in detail in accordance with other Classes appropriate to the work.

EXAMPLES

Two examples follow:

Measured Example (IE.1)

The Measured Example IE.1, covers the taking of the dimensions for the pipeline work from MH.1 to MH.3 shown on Drawing No. I/D/1. This is considered to be a pipe run of particular cost characteristic. The dimensions for other pipe runs would be recorded in a similar way. In the Example a schedule is used in combination with a conventional dimension sheet for the taking off. The work on the schedule being totalled for billing direct. The work on the dimension sheet being that which would be collected on to an abstract with that related to other pipe runs.

Specimen Bill of Quantities, Example (IE.2)

The specimen Bill at Example IE.2 is for the Work Items (excluding the General Items) for the Surface Water Sewers, excluding the Pumping Station and Rising Main, shown on Drawing No. I/D/1. In keeping with Note I2 of the CESMM, the Bill indicates the location of the runs of pipes and groups together those considered to involve similar cost characteristics.

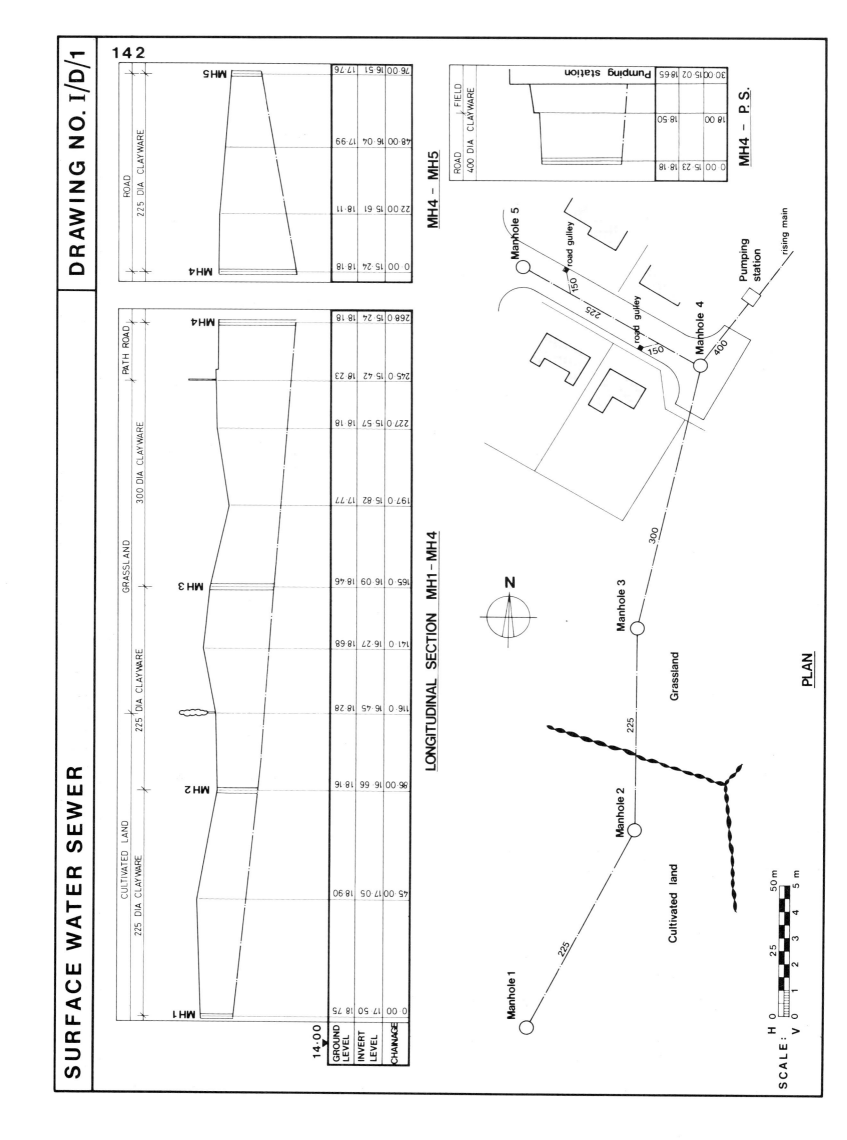

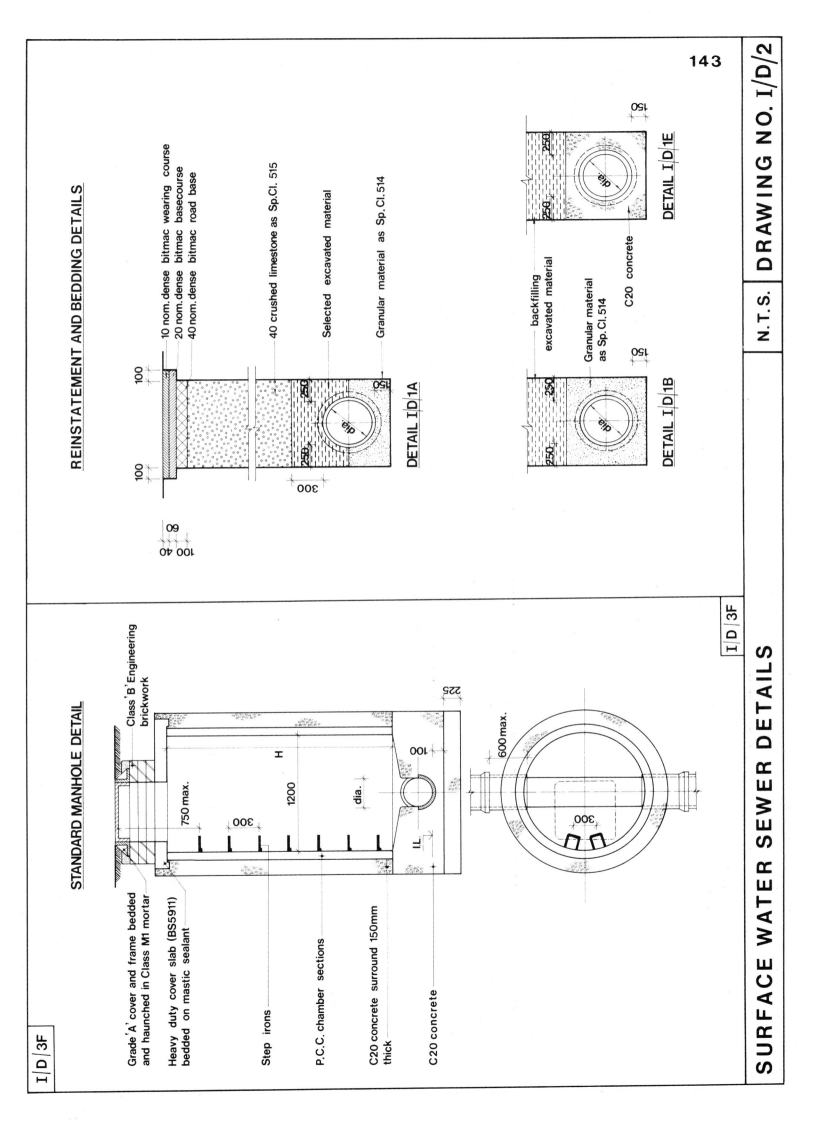

I|D|3F

REINSTATEMENT AND BEDDING DETAILS

10 nom. dense bitmac wearing course
20 nom. dense bitmac basecourse
40 nom. dense bitmac road base

40 crushed limestone as Sp.Cl. 515

Selected excavated material

Granular material as Sp.Cl.514

DETAIL I|D|1A

DETAIL I|D|1E

backfilling
excavated material

Granular material
as Sp.Cl.514

C20 concrete

DETAIL I|D|1B

STANDARD MANHOLE DETAIL

Class 'B' Engineering
brickwork

Grade 'A' cover and frame bedded
and haunched in Class M1 mortar

Heavy duty cover slab (BS5911)
bedded on mastic sealant

Step irons

P.C.C. chamber sections

C20 concrete surround 150mm
thick

C20 concrete

750 max.

1200

dia.

I.L.

600 max.

I|D|3F

N.T.S. DRAWING NO. I/D/2

SURFACE WATER SEWER DETAILS

		PIPES				
SURFACE WATER SEWER – DRG No I/1		225 ∅ Vit. Clay Pipes flex. Jts. in Trenches				
SCHEDULE No 1.						
MH 1 – MH2 – MH3		n.e 1	1 – 1.5	1.5 – 2	2 – 3	3 – 4
MH 1 – Chain 45 (45.00)	45.00					
GL Inv Depth	18.75					
MH1 18.75 – 17.50 = 1.25 | 1.50	26.25			26.25		
Ch.45 18.90 – 17.05 = 1.85 | 1.25						
Diff 0.60 | 0.25	18.75					
$\dfrac{0.25 \times 45.00}{0.60}$ 18.75 Less ½/MH 1	0.60					
	18.15		18.15			
Chain 45 – MH2 (86 – 45 = 41.00)	41.00					
GL Inv Depth Less ½/MH 2	0.60					
Ch.45 1.85	40.40			40.40		
MH.2 18.16 – 16.66 = 1.50						
MH 2 – Chain 116 (116 – 86 = 30.00)	30.00					
GL Inv Depth Less ½/MH 2	0.60					
MH 2 1.50	29.40			29.40		
Ch 116 18.28 – 16.45 = 1.83						
Chain 116 – 141 (141 – 116 = 25.00)	25.00					
GL Inv Depth	7.33			7.33		
Ch.116 1.83 | 2.00	17.67				17.67	
Ch 141 18.68 – 16.27 = 2.41 | 1.83						
Diff. 0.58 | 0.17						
$\dfrac{0.17 \times 25.00}{0.58} = 7.33$						
Chain 141 – MH 3 (165 – 141 = 24.00)	24.00					
GL Inv Depth Less ½/MH 3	0.60					
Ch.141 2.41	23.40			23.40		
MH 3 18.46 – 16.09 2.37						
			18.15	103.38	41.07	

SURFACE WATER SEWER DRG.No.I/D/1

EXAMPLE IE.1

Pipes MH 1-2-3

NOTE — Bill direct from Schedule No.1 pipes in trenches MH 1-2-3

Ancills

1 — Crossings, hedge, pipe nom. bore 225 mm.
(K641

½/MH 1
```
        116.00
          0.83
        116.83
```

116.83 — Reinstatement cultiv. land, pipe nom. bore 225 mm   (K781
```
165.00
116.00
 49.00
```

49.00 — Reinstatement grass-land, pipe nom. bore 225 mm   (K751
```
Less MHs ²/1.65 =  165.00
                     3.30
                   161.70
```

Supports etc.

Beds, haunches + surrs. in gran. matl. type 2 Spec. clause 514

161.70 — Beds depth 150 mm & surrounds to pipes nom. bore 225 mm as Drg.No I/D/8 (L932.3

```
                    86.00
2/½/1.65 = 1.65
                    84.35

                   116.00
                    86.00
                    30.00
½/ 1.65 =  0.83
MH2-Ch116 =  29.17

225    0.60  0.20
500    0.20  0.00
725   2)0.80  0.20
        0.40  0.10
```

Ancills. to laying

Extras to excavn. and backfill

```
84.35
 0.73
 0.40

29.17
 0.73
 0.10
```

{ Excavn of rock; trenches   (L110.1

COMMENTARY

The calculations on the left hand side of Schedule No. 1, determine by interpolation the points of division between the different depth ranges of the pipe trenches.

The linear dimension for reinstatement includes the lengths occupied by manholes. See Table 9.06 and associated Commentary.

The linear dimension for granular beds, haunches and surrounds, excludes the lengths occupied by manholes. See Table 9.10.

Rock contours are not shown on the Drawing. The Example assumes that rock will be encountered at 600 mm above the base of the granular bed at Manhole No. 1 and will run out to zero at Chainage 116. The width entered for excavation of rock is determined as provided in Note L2 of CESMM. See Commentary following Table 9.08.

PART 3 : SURFACE WATER SEWER

Number	Item Description	Unit	Quantity	Rate	Amount	
					£	p
	PIPEWORK - PIPES					
	Vitrified clay pipes to BS.65 and 540: Part 1, "Extra strength" with flexible joints to BS.65 and 540: Part 2					
	Between manholes 1 and 2 and 2 and 3 Drawing No. I/D/1					
I123.1	Nominal bore 225 mm, in trenches, depth 1 - 1.5 m.	m	18			
I124.1	Nominal bore 225 mm, in trenches, depth 1.5 - 2 m.	m	103			
I125.1	Nominal bore 225 mm, in trenches, depth 2 - 3 m.	m	41			
	Between manholes 3 and 4 and 4 and Pumping Station, Drawing No. I/D/1					
I124.2	Nominal bore 300 mm, in trenches, depth 1.5 - 2m.	m	6			
I125.2	Nominal bore 300 mm, in trenches, depth 2 - 3 m.	m	96			
I135	Nominal bore 400 mm, in trenches, depth 2 - 3 m.	m	2			
I136	Nominal bore 400 mm, in trenches, depth 3 - 4 m.	m	27			
	Between manholes 4 and 5, Drawing No. I/D/1					
I123.2	Nominal bore 225 mm, in trenches, depth 1 - 1.5 m.	m	9			
I124.3	Nominal bore 225 mm, in trenches, depth 1.5 - 2 m.	m	20			
I125.3	Nominal bore 225 mm, in trenches, depth 2 - 3 m.	m	45			
	(1)			Page Total		

Number	Item Description	Unit	Quantity	Rate	Amount	
					£	p
	PIPEWORK – PIPES					
	Vitrified clay pipes to BS.65 and 540:Part 1, "Extra strength" with flexible joints to BS.65 and 540: Part 2					
	Road Gully branches					
I112	Nominal bore 150 mm, in trenches, depth not exceeding 1 m.	m	14			
I113	Nominal bore 150 mm, in trenches, depth 1 – 1.5 m.	m	3			
I114	Nominal bore 150 mm, in trenches, depth 1.5 – 2 m.	m	3			
	PIPEWORK – FITTINGS AND VALVES					
	Vitrified clay pipe fittings to BS.65 and 540:Part 1, "Extra strength" with flexible joints to BS.65 and 540: Part 2					
J111	Bends, nominal bore 150 mm.	nr	4			
J122	Junction, single "Y", nominal bore 225 mm.	nr	2			
	PIPEWORK – MANHOLES AND PIPEWORK ANCILLARIES					
	Manholes as Drawing No. I/D/3F the chamber sections surrounded with concrete as Specification clause 503, with Grade A heavy duty coated cast iron covers and frames, 600 mm square clear opening, as BS.497					
K151	Precast concrete, depth 1.35 m.	nr	2			
K152	Precast concrete, depth 1.60 m.	nr	1			
K153	Precast concrete, depth 2.47 m.	nr	1			
K154	Precast concrete, depth 3.04 m.	nr	1			
	(2)			Page Total		

Number	Item Description	Unit	Quantity	Rate	Amount	
					£	p
	PIPEWORK - MANHOLES AND PIPEWORK ANCILLARIES					
	Vitrified clay gullies and bedding and surrounding with concrete as Specification clause 508					
K320	400 mm Diameter road gully, 750 mm deep, with rodding eye, stopper and 150 mm outlet and coupling joint and with 450 mm nominal width cast iron coated hinged grating, as BS. 497 reference GA2-450 on two courses of one brick thick Class B engineering brickwork raised off gully surround.	nr	2			
	Crossings					
K641	Hedge, pipe nominal bore 225 mm.	nr	1			
K662	Fence, pipe nominal bore 250 - 500 mm.	nr	1			
	Reinstatement					
K731	Breaking up and temporary and permanent reinstatement of roads, as detail Drawing No. I/D/1A, pipe nominal bore not exceeding 250 mm.	m	97			
K732	Breaking up and temporary and permanent reinstatement of roads as detail Drawing No. I/D/1A, pipe nominal bore 250 - 500 mm.	m	37			
K742	Breaking up and temporary and permanent reinstatement of footpaths, as detail Drawing No. I/D/1A, pipe nominal bore 250 - 500 mm.	m	4			
K751	Reinstatement of grassland, pipe nominal bore 225 mm.	m	49			
K752	Reinstatement of grassland, pipe nominal bore 250 - 500 mm.	m	92			
K781	Reinstatement of cultivated land, pipe nominal bore 225 mm.	m	117			
	(3)			Page Total		

Number	Item Description	Unit	Quantity	Rate	Amount	
					£	p
	PIPEWORK – SUPPORTS AND PROTECTION, ANCILLARIES TO LAYING AND EXCAVATION					
	Extras to excavation and backfilling					
L110.1	Excavation of rock; trenches.	m3	27			
L110.2	Excavation of rock; manholes.	m3	2			
L160	Backfilling above the Final Surface with 40 mm nominal maximum size broken limestone as Specification clause 515; trenches in roads and footpaths.	m3	177			
	Beds, depth 150 mm, haunches and surrounds in granular material as Specification clause 514					
L931.1	Beds and haunches to pipes nominal bore 150 mm, as Drawing No. I/D/1A.	m	6			
L932.1	Beds and haunches to pipes nominal bore 225 mm, as Drawing No. I/D/1A.	m	74			
L932.2	Beds and haunches to pipes nominal bore 300 mm, as Drawing No. I/D/1A.	m	22			
L933.1	Beds and haunches to pipes nominal bore 400 mm, as Drawing No. I/D/1A.	m	17			
L932.3	Beds and surrounds to pipes nominal bore 225 mm, as Drawing No. I/D/1B.	m	162			
L932.4	Beds and surrounds to pipes nominal bore 300 mm, as Drawing No. I/D/1B.	m	79			
L933.2	Beds and surrounds to pipes nominal bore 400 mm, as Drawing No. I/D/1B.	m	12			
	Beds, depth 150 mm, and surrounds as detail Drawing No. I/D/1E, mass concrete as Specification clause 503					
L941	Beds and surrounds to pipes nominal bore 150 mm.	m	14			
	(4)			Page total		

10 Structural Metalwork and Miscellaneous Metalwork—CESMM Classes: M and N

The CESMM provides two main classifications for metalwork and designates them Classes M and N. Class M features components of structural metalwork and sets the rules for their measurement. Class N is appropriate for the measurement of metalwork components, not included in Class M, which are associated with metalwork for structures and also for various other metalwork components which may be encountered in civil engineering work. The application of each Class is subject to the specific exclusions given in the "Excludes" preceding the classification table for each Class in the CESMM. References to work included in Classes M or N are made in the "Excludes" above the classification tables in Classes P, Q, S and V.

STRUCTURAL METALWORK - CESMM CLASS M

The method of measurement set out in the CESMM, relies on drawings to provide some of the information which tenderers require. For structural metalwork it is usual to provide tenderers with dimensioned drawings showing lay out, sections and sizes of the members and details of any connections and other fittings.

Table 10.01 Structural Metalwork

Generally – The mass of members shall be that of the plates, rolled sections, shear connectors, stiffeners, cleats, packs, splice plates and other fittings (Note M5). Exclude mass of weld fillets, bolts, nuts, washers and protective coatings (Note M5).

Calculate the mass of members from the overall length of the members with no deduction for splay or mitred ends (Note M4). Make no deduction from the mass for notches and holes less than 0.1 m2 in area (Note M5). Make no allowance for rolling margin or other permissible deviations (Note M5).

Take mass of mild steel grades 43A1 and 43A as 785 kg/m2 per 100 mm thickness (7.85 t/m3). Take mass of other metals as stated in the Specification or the suppliers catalogue (Note M3).

Identify temporary structural metalwork (Note M2).

Include rails for overhead cranes in this Class. State where fixing clips and resilient pads are used to secure crane rails (Note M6).

Classify under Class N other metal components not included in this Class but associated with metal structures (Note M6).

COMMENTARY

Structural Metalwork (refer to Table 10.01)

Table 10.01 outlines the Notes in Class M of the CESMM which are applicable to structural metalwork generally. Quantities attached to the members give the calculated mass of the members. The mass of a member includes the mass of attached fittings. Fabrication of structural metalwork is given separately from erection, see subsequent Tables and Commentary. The Notes mentioned in Table 10.01 regarding mass apply to both fabrication and erection.

Quantities are given in tonnes. They may be entered in the Bill of Quantities to the nearest one tenth of a tonne. See Paragraph 5.18 of the CESMM.

Table 10.02   Fabrication and Erection

Fabrication

1st Division		2nd Division		3rd Division

Fabrication of members for bridges	State materials and grades of materials (Note M1)
	
	Identify tapered and castellated members (Note M2)
		Main members:
		  Rolled sections t
		  Plates and flats t
		  Built-up box or hollow sections t
		Subsidiary members:
		  Deck panels t
		Subsidiary members:
		  Bracings t
		  External Diaphragms t
				Straight on plan
				
				Curved on plan
				
				Straight on plan and cambered
				
				Curved on plan and cambered

Fabrication of members for frames Fabrication of other members	State materials and grades of materials (Note M1) Identify tapered and castellated members (Note M2)	Columns t Beams t Trestles, towers and built-up columns t Trusses and built-up girders t	State details of members comprising boom and infill construction (Note M7)	Straight on plan Curved on plan Straight on plan and cambered Curved on plan and cambered
		Bracings t Purlins and cladding rails t		
		Grillages t		
		Anchorages and holding down bolt assemblies nr	Measure by number of complete assemblies (Note M10) State particulars of types of anchorages or assemblies (Note M10)	

Erection

1st Division		2nd Division		3rd Division
Erection of members for bridges Erection of members for frames Erection of other members	Identify and locate separate bridges and structural frames and where appropriate, parts of bridges or frames (Note M8)	Trial erection t Permanent erection t		
		Site bolts nr	State type and enumerate as 2nd Division features	State diameter range as 3rd Division features

COMMENTARY

Structural Metalwork (cont.

Note M2 of the CESMM requires that item descriptions identify temporary structural metalwork. The Note is applicable to temporary structural metalwork specifically required by the Contract and does not imply that it is necessary to measure any temporary work the Contractor may require.

In order to comply with Note M6, overhead crane rails with fixing clips and resilient pads are given separately using non-standard 2nd Division descriptive features.

Metalwork components associated with metal structures but not listed in the Class M classification table are measured and classified under Class N.

Fabrication

For the purpose of measurement, the fabrication of structural metalwork is divided into three main classifications, i.e. for bridges, for frames and for other members. Within each of the main classifications members are classified and described in accordance with 2nd Division features which identify the members. The identified members are further classified according to their shape as set out in 3rd Division features. Item descriptions identify members which are tapered or which are castellated. Cranked or other unusual members are similarly identified.

The quantities for fabrication of structural metalwork give the mass in tonnes for each classification, calculated as outlined in Table 10.01. See also preceding Commentary "Structural Metalwork".

Note M7 of the CESMM requires item descriptions for trestles, towers and built-up columns and for trusses and built-up girders to state details of the members which comprise the boom and infill construction. The sizes and sections of these members are given in the item descriptions.

Separate items are given for anchorage and holding down bolt assemblies. Items are given to cover the complete assemblies. Each type is described or identified and quantities give the number of each. Bridge bearings are measured and classified as provided in Class N.

Erection

Items for the erection of structural metalwork group together the mass of the members in one item for each bridge or structural frame (or where appropriate each part of a bridge or structural frame) which is identified and located in the item description. Quantities give the mass in tonnes for each bridge, frame or part as the case may be. Separate items are given for site bolts, see Table 10.02. Trial erection is given separately where this is specifically required. It is appropriate for different parts of bridges and structural frames to be given as separate items, rather than as one item for the whole, where for identification purposes it would be helpful to do so, or to separate parts considered to have different cost characteristics.

Table 10.03 Surface Treatment

1st Division		2nd Division	3rd Division
Surface treatment m2	Applies to treatment carried out before delivery to Site Class surface treatment on Site after erection as painting (Note M9)	Blast cleaning Pickling Flame cleaning Wire brushing Metal spraying Galvanising Painting	

COMMENTARY

Surface Treatment (refer to Table 10.03)

The surface area of structural metalwork required to be treated before delivery to the site is measured in m2 and separate items are given for each type of treatment.

MISCELLANEOUS METALWORK — CESMM CLASS N

Table 10.04   Miscellaneous Metalwork

Generally — Separate items are not required for:-
      (i) erection and fixing or for the provision of fixings (Note N4), or for
      (ii) the components of assemblies (Note N2)

No deduction from masses or areas for openings and holes each less than 0.5m2 in area (Note N6)

Use 1st Division item codes 3 — 8 for miscellaneous metalwork not listed. Generally the units of measurement for such work shall be the tonne. (Note N3)

1st Division	2nd Division			3rd Division
	Stairways and landings	t	Include mass of all components and attached pieces (Note N2)	State specification and thickness of metal used. State off site surface treatment. State principal dimensions, or as an alternative to the foregoing identify by mark numbers related to details shown on drawings or given in the Specification (Note N1)	Refer to the classification table for 3rd Division features.
	Walkways and platforms	t			
	Ladders	m			
	Handrails	m	Measure along top member (Note N5)		
	Bridge parapets	m			
	Rectangular frames	m	Measure along external perimeter (Note N5)		
	Plate flooring	m2			
	Open grid flooring	m2			
	Welded mesh panelling	m2			
	Duct covers	m2			
	Tie rods	nr			
	Walings	nr			
	Bridge bearings	nr			
	Tanks	nr			
	Tank covers of stated area	nr			

COMMENTARY

Miscellaneous Metalwork (refer to Table 10.04)

In addition to metalwork components for which classifications are listed in Class N, the Class provides classification numbers and measurement units for unlisted metalwork components. The unused item Codes beyond N2 are used in conjunction with ad-hoc descriptive features, compiled by the person preparing the Bill, for metalwork components not listed in Class N or catered for in other Classes.

The information to be given in the descriptions for miscellaneous metalwork items is set out in Notes N1 and N2 of the CESMM. It is considered preferable to use the alternative permitted by Note N1 and for brief identifying descriptions to make reference to drawings and the specification from which can be obtained the details which otherwise must be stated in the item descriptions. The items for miscellaneous metalwork are to be taken to include erection, fixing, provision of fixings and component assemblies to the extent specified or otherwise indicated. See "Generally" Table 10.04. Where item quantities are given by mass they include the mass of all components and attached pieces.

EXAMPLE ME 1.

Measured Example

The Measured Example is for part of the work for a steel bridge over a cutting. Excavation, foundations, abutments and other work which forms part of the bridge are not measured in the Example. The dimensions are for the structural metalwork and related incidental items only.

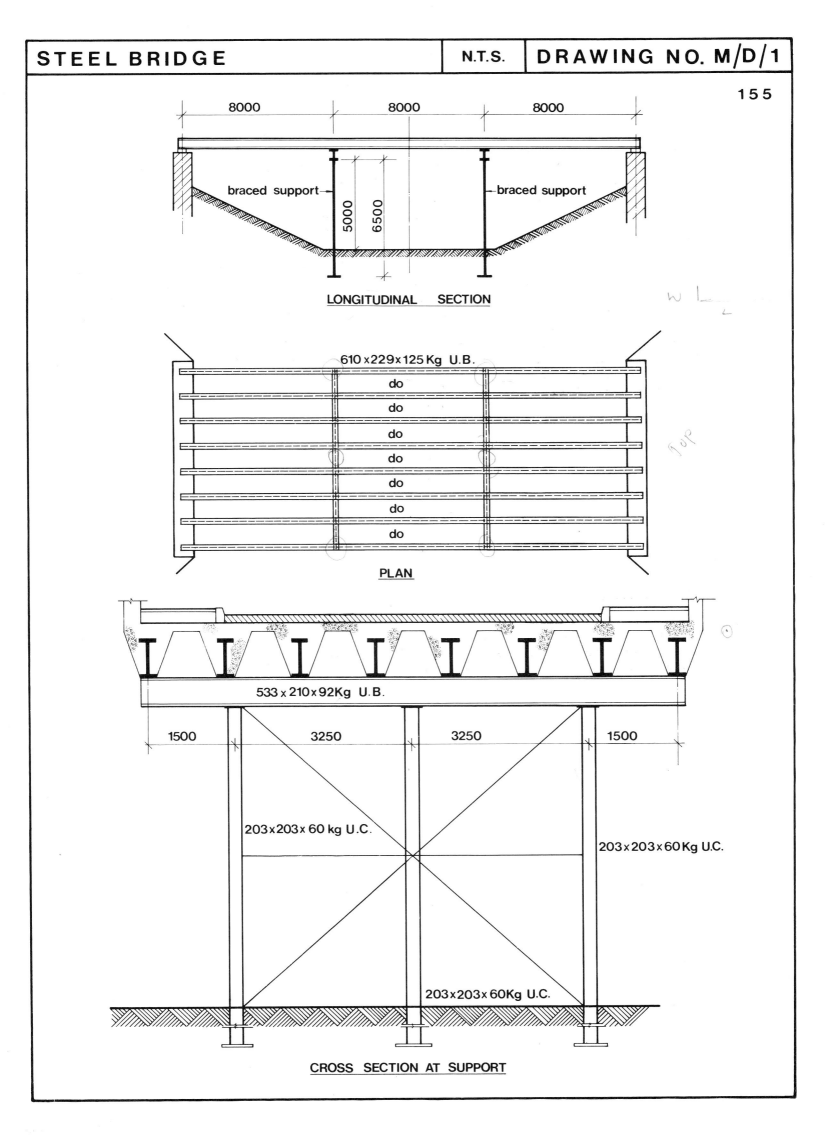

8000   8000   8000

braced support   5000   6500   braced support

LONGITUDINAL SECTION

610 × 229 × 125 Kg U.B.

do
do
do
do
do
do
do

PLAN

533 × 210 × 92Kg U.B.

1500   3250   3250   1500

203 × 203 × 60 kg U.C.

203 × 203 × 60 Kg U.C.

203 × 203 × 60Kg U.C.

CROSS SECTION AT SUPPORT

156

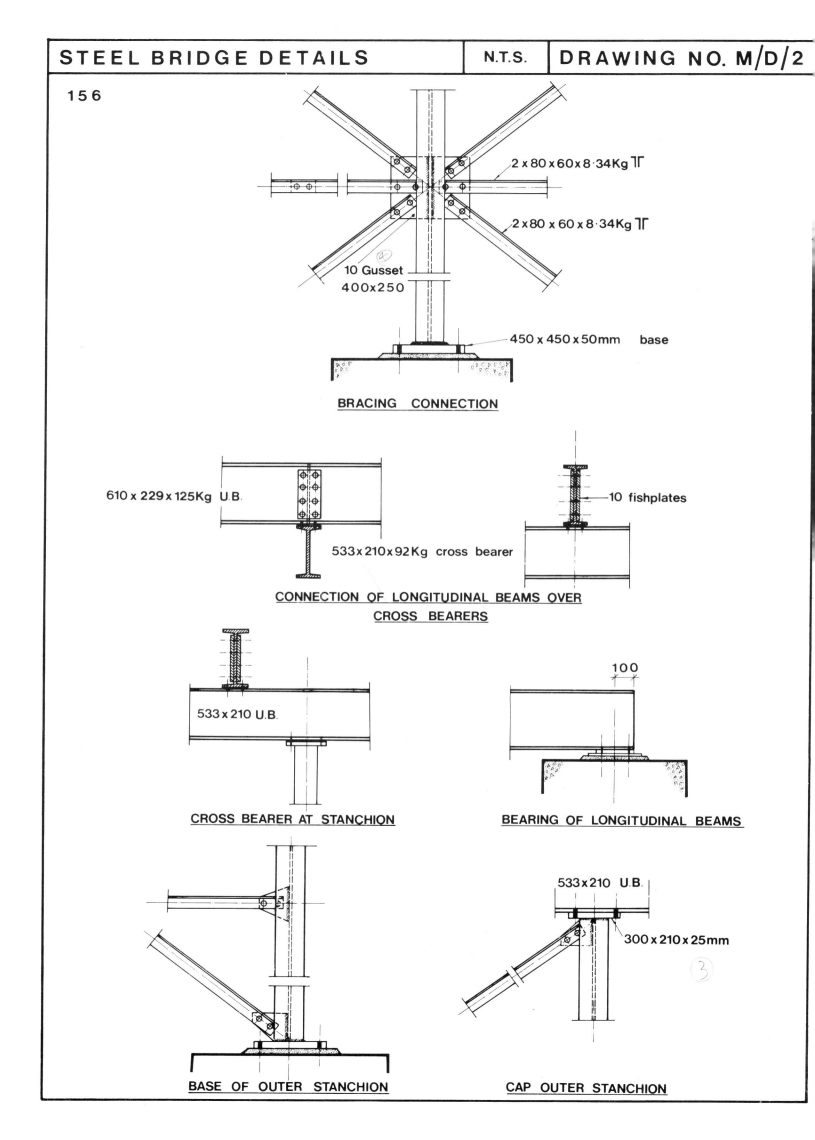

2 x 80 x 60 x 8·34 Kg ⊤⊤

2 x 80 x 60 x 8·34 Kg ⊤⊤

10 Gusset
400 x 250

450 x 450 x 50mm   base

BRACING   CONNECTION

610 x 229 x 125 Kg U.B.

10 fishplates

533 x 210 x 92 Kg cross bearer

**CONNECTION OF LONGITUDINAL BEAMS OVER
CROSS BEARERS**

533 x 210 U.B.

100

CROSS BEARER AT STANCHION

BEARING OF LONGITUDINAL BEAMS

533 x 210 U.B.

300 x 210 x 25mm

BASE OF OUTER STANCHION

CAP OUTER STANCHION

STEEL BRIDGE OVER CUTTING . DRGs M/D/1 + 2

<u>All steel Grade 43 A unless</u>
<u>o/w described</u>

Deck beams
3/8000   24000
2/ 100   200
            24200

		Kg.		
8/	24.20 × 125.00	24200	Fabricn. of membs. for bridges, main membs., rolld. sectns. stra. on plan (M111 (610 x 229 U.B. (200 x 10 fish plate +	
8/2/2/	0.55 × 15.70	276·32		
		24476·32		
			Erection of membs for bridges, perm. erection. (M420	
8/2/	8	128 00	Site bolts, black, diam. 20 - 24 mm (M433	
8/2/	4	64 00		
		192·00		
8/2/	1	16·00	Bridge beargs, slide; inc. plates + anchor bolts as detail Drg. M/D/2 (N252 + Conc. access _ inserts, fix and cast in only set 4 anchor bolts + grout beneath bearg. plate as detail Drg M/D/2 (G832.1	

(1)

COMMENTARY

Items for the fabrication of members include the provision of the structural metalwork components. They are given separately from the items for erection. See Table 10.02 and Commentary given previously. The dimensions for the fabrication and erection items are set down in the Example in a manner which allows the mass in kg. to be calculated on the dimension sheets. The first dimension of each set of two is the length of the member or attachment in metres, the second is its mass in kg. per metre.

When billing the structural metalwork, appropriate items can be listed under the sub headings "Fabrication of members straight on plan" and "Erection of members permanent erection" to avoid having to repeat these phrases in various item descriptions.

The code attached to the item for site bolts indicates that it relates to the erection of members for bridges. The item description given in the Bill for the bolts must state or otherwise indicate that the bolts relate to the erection of members for bridges. The site bolts here measured are those for the fish-plate connections and those to attach the bridge beams to the trestle cross members.

Additional description is given in the items of bridge bearings and inserts to make clear the extent of the work intended to be included in the items.

STEEL BRIDGE OVER CUTTING _ DRGs M/D/1 & 2

			Beams (cont.
8/	24.20		{Surf. treatment, blast
	2.08	?402.688?	{cleang. Spec. clause 'x'
8/2/2/	1.20		(M710
	0.42		&
			{Surf. treatment, 2 c.
			{calcium plumb. primer
			(M770
			Trestles.
			{Fabricn. of membs.
			{for frames, trestles
			{stra. on plan, compris
			ing. 203×203×60 Kg/m
			U.C. legs, 533×210×92
			Kg/m cross membs,
			80×60×8.34 Kg/m
2/3/	6.50	Kg.	angle bracg & 10 mm
	×		p.lates (M231
	60.00	✓	
2/3/	0.45		(203×203 U.C. (legs
	×		
	177.00	✓	(450 × 50 p.late (base
2/3/	0.21	✓	
	×		(300 × 25 plate (cap
	58.90		
2/2/	0.40		(250 × 10 plate (gussets
	×	V	
2/2/3/	19.60		(" " " ("
	0.20	?	
	×		
	19.60		
2/	9.73	?	(533 × 210 U.B (cross
	×		(memb
	92.00		
2/2/2/	3.25	✓	(80 × 60 L (horiz
	×		
	8.34		(80 × 60 L (braces
2/4/2/	4.60		
	×		(80 × 10 plate (packers
	8.34		&
2/6/	0.15		
	×		{Erection of membs.
	6.28		{for frames, perm.
			{erection (M520

(2)

COMMENTARY

Surface treatment measured under CESMM Class M is that required to be carried out before the structural metalwork is delivered to the Site. On Site surface treatment before or after erection is classed as painting under CESMM class V.

The second dimension of each set of two here given for surface treatment is the surface area in m2 per metre linear of the member. It is taken from a Steel Sections handbook.

Details of the members are stated in the item descriptions for trestles as provided in CESMM Note M7.

When taking dimensions for structural metalwork, it is convenient to follow the dimensions of a main member with those for the plates and fittings attached to it before proceeding to set down the dimensions of another main member.

STEEL BRIDGE OVER CUTTING - DRGs M/D/1+2

Trestles (cont

2/2/3/	2	{ Site bolts, black diam
2/	12	{ 20 - 24 mm (M533
2/6/	2	(packers.
2/3/	4	

2/ 3

Fabricn. of membs for fras., anchorage & h.d. bolt assemblies, 4 No h.d. bolts with p.late washers as detail Drg M/D/2 (M280

&

Conc. access., inserts, fix and cast in only set 4 h.d. bolts & grout beneath base plate. Drg M/D/2   (G832.2

2/3/	6.50
	1.20
2/3/	0.55
	0.55
2/2/3/	0.24
	0.33
2/2/2/	0.41
	0.26
2/2/2/3/	0.21
	0.26
2/	9.73
	1.84
2/2/2/	3.25
	0.28
2/4/2/	4.60
	0.28
2/2/6/	0.16
	0.09

{ Surf. treatment, blast clean. a.b. (M710

&

Surf treatment. 2c. primer. a.b. (M770

(3)

11 Timber—CESMM Class: O

Provision for the classification of formwork, timber piles, timber sleepers and timber fencing is made in other Classes of the CESMM, and these are specifically excluded from Class O. See the "Excludes" at the head of the classification table of Class O in the CESMM. The CESMM is not intended to apply to the measurement of Building work and the further specific exclusion of building carpentry and joinery makes it clear that Class O is intended to apply to timber in works of civil engineering construction.

Cross reference is made to work included in Class O in the "Excludes" at the head of the classification table in Class N of the CESMM.

Table 11.01 Timber Components, Timber Decking and Fittings and Fastenings

Generally   Separate items are not required for fixing or for boring and cutting (Note 07)

For timber components in buildings, refer to Note 01

1st Division		2nd Division	3rd Division
Hardwood components for general use m Hardwood components for marine use m Softwood components m	Measure overall length with no deduction for joints (Note 04) State structural use and location of components longer than 3 m (Note 03) State species (Note 02) State any impregnation requirement (Note 02) State any special surface finish (Note 02)	State nominal gross cross-sectional dimensions (Note 02) Use nominal gross cross-sectional area for classification (Note 05)	State length range as 3rd Division features, or state actual length where only one in one item.
Hardwood decking for general use m2 Hardwood decking for marine use m2 Softwood decking m2	State species (Note 02) State any impregnation requirement (Note 02) State any special surface finish (Note 02) No deduction for openings and holes each less than 0.5 m2 in area (Note 08)	State nominal gross thickness (Note 02) Use nominal gross thickness for classification (Note 06)	
Fittings and fastenings nr		Straps	
		Spikes	
		Coach screws	
		Bolts	
		Plates	

COMMENTARY

Class 0 (refer to Table 11.01)

Item descriptions for timber components and timber decking state the timber species and the grade. They state also any impregnation requirements and any special surface finish.

Separate items are not required for fixing timber components and decking, or for any boring or cutting of the timber necessary for their fabrication and fixing. Item descriptions for otherwise similar components or decking distinguish different methods of fixing where the methods are adjudged to have cost differences.

Timber Components

Timber components are measured by length in metres. Length is measured as illustrated in Figure 01. below.

The cross-sectional dimensions of the timber components before reduction by any surface finish (i.e. the nominal gross cross-sectional dimensions) are stated in item descriptions. These dimensions are used also to calculate the nominal gross cross-sectional area for Second Division classification purposes.

Item descriptions for timber components longer than 3 m, state the location and the structural use of the components.

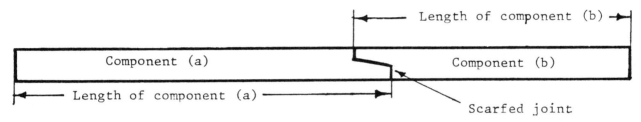

Fig. 01. Application of Note 04 of CESMM to measurement of timber components.

Timber Decking

Timber decking is measured by area and is given in m2. The area measured includes the areas of any openings individually less than 0.5 m2.

The spaces between the boards or planks in open spaced decking is deducted where the area of individual spaces is 0.5 m2 or more. Where individual spaces are less than 0.5 m2 in area the work is measured overall the spaces. It is usual for item descriptions for open spaced decking to state whether or not the spaces have been deducted.

The nominal gross thickness of decking is stated in item descriptions. The Second Division thickness range is overridden by Note 02. The widths of the boards or planks used for the decking are stated or otherwise identified in item descriptions.

Fittings and Fastenings

Fittings and fastenings, of the type noted in the 2nd Division of this feature in Table 11.01, which are used to connect together or secure timber components or decking are given as separate items. The quantities give the number of each type. Fittings and fastenings other than those listed, such as timber connectors, etc., qualify also to be measured as separate items. Additional description or drawing or specification reference will usually be attached to the standard descriptive features to identify materials, types, sizes and ancillary components, such as nuts, washers, etc., which are intended to be included in the items.

Separate items are not required for fixing fittings and fastenings or for the cutting or boring of the timber necessary for their fixing.

Class O (cont.

Specimen Item Descriptions for Timber Components and Decking

Number	Item description	Unit
	TIMBER	
	Softwood components	
	Douglas fir, impregnated with preservative, wrought finish	
0311	100 x 100 mm, length not exceeding 1.5 m.	m
0313	100 x 100 mm, length 3 - 5 m, handrail, twice rounded.	m
	TIMBER - Work affected by tidal water, high water level ordinary spring tides assumed 1.98 m above ordnance datum, low water level ordinary spring tides assumed 1.83 m below ordnance datum	
	Hardwood components for marine use	
	Greenheart fixed at positions affected between high and low tides	
0242.1	200 x 300 mm, length 1.5 - 3 m.	m
0242.2	300 x 300 mm, length 1.5 - 3 m.	m
	Greenheart fixed at positions affected at all states of the tide	
0243	300 x 325 mm, length 3 - 5 m, diagonal braces between piles.	m
0244	300 x 300 mm, length 5 - 8 m, horizontal deck bearers between piles.	m
0254	300 x 400 mm, length 5 - 8 m, bearers to support braces between piles.	m
	Hardwood decking for marine use	
	Greenheart, wrought finish, fixed at positions affected at all states of the tide	
0540.1	Thickness 100 mm, plank widths 300 mm.	m2
0540.2	Thickness 100 mm, plank widths 300 mm, with 19 mm open spaces between (no deduction for spaces).	m2

COMMENTARY

The cross-sectional dimensions of timber components are stated in item descriptions (CESMM Note 02). To state also cross-sectional area is considered unnecessary duplication of information. The standard Second Division features are, therefore, omitted from the descriptions. The features serve as an instrument for coding.

See Commentary on Paragraph 5.20 in Chapter 2 regarding "Work affected by bodies of water".

12 Piles and Piling Ancillaries—CESMM Classes: P and Q

Two Classes of the Work Classification are allocated in the CESMM to piling. Class P applies to the piles and to the driving or boring. Work to the piles other than that covered by Class P and work associated with piling is measured, to the extent that it is expressly required, under Class Q. Boring for site investigation, ground anchors and walings and tie rods are listed in the CESMM as specific exclusions from these Classes.

PILES — CESMM CLASS P

Items for temporary works and plant are not given in the Bill of Quantities unless specific requirements are specified for them. Features of classification for piling plant (Items, Codes A337 and A338) and for temporary works (Item, Code A363) are listed in Class A of the CESMM. They serve to remind that tenderers may insert items for these, if they so choose, as Method-Related Charges.

Class P requires a set of items to be given for each group of piles. A group of piles being regarded as all piles of the same material, the same type and of the same cross-section installed in one location on the Site.

Table 12.01   Piles Generally

State materials of which piles are composed (Note P1).

The Commencing Surface is that at which boring or driving is anticipated to begin (Note P6).

COMMENTARY

Piles Generally (Refer to Table 12.01)

Reference is made in Table 12.01 to Notes in Class P which are applicable to all types of piles.

Item descriptions for piles state the materials of which the piles are composed. They state also the actual section characteristics of the piles as provided in Note P1 of the CESMM. The Note overrides the Second Division ranges given in the classification table of Class P of the CESMM.

Commencing Surface

For piling on land the Commencing Surface may be the Original Surface or a surface below or above the Original Surface. For work in water the Original Surface is the natural bed of the river, sea, etc., the Commencing Surface may be the Original Surface or another surface. For example, the bed left after dredging. In all cases, item descriptions for driving and boring must identify the Commencing Surface where it is not the Original Surface (See Paragraph 5.21 of the CESMM). Bored and driven depths are measured from the Commencing Surface as provided in Note P6 of the CESMM.

Table 12.02   Cast in Place Piles

1st Division			2nd Division
Bored cast in place concrete piles		Identify preliminary and contiguous bored piles (Note P11)	State cross-sectional dimensions or nominal diameter (Note P1)
Driven cast in place piles	For piles comprising light steel casings filled with concrete, see Note Q5		

3rd Division		
The following separate items, each indicated by a bracketed letter below, are required for each group of piles		
(a) Number of piles	nr	Give item for the number of piles in the group (Note P3)
(b) Length of piles	m	Give item for total concreted length of the piles in the group (Note P3), including the length of enlarged bases but excluding any cut-off tolerances (Note P7)
(c) Depth bored or driven	m	Give items for total depth bored or driven for the group of piles divided into the ranges of depth given in the Third Division (Note P3). Include the whole depth of boring or driving for each pile in the item in the range of which the depth occurs (Note P8). Measure along axes of the piles from the Commencing Surface to the bottom of the shafts of bored piles and to the bottom of the casing for driven cast in place piles (Note P6). Where piles are raked, state the angle of rake within ranged increments of 10 degrees (Note P11)

COMMENTARY

Cast in Place Piles (Refer to Table 12.02)

A set of several different items are given for each group of cast in place piles. In the Bill of Quantities an item giving the number of piles in a group is followed by separate items for the length of piles and for the depth bored or driven for the group. The latter giving total depth divided into depth ranges as noted against feature (c) in Table 12.02. A set of items following the same format is given for each group. (See Table 12.02 which sets out the descriptive features and notes also measurement conventions and units of measurement).

Piles Concreted Above Cut-off Level

Where piles are expressly required to be concreted to a higher level than cut-off level, for subsequent removal, the additional length is measured as length of piles. It is given separately where it is of a different grade of concrete from that in the piles. Subsequent removal of the additional length is classified under Class Q as "preparing heads" and is made the subject of items which identify the work required and give the number of the additional lengths to be removed.

Table 12.03 Preformed Concrete, Timber and Isolated Steel Piles

1st Division		2nd Division
Preformed concrete piles	Identify preliminary piles (Note P11)	State cross-sectional or nominal diameter dimension
Preformed prestressed concrete piles		
Preformed concrete sheet piles		
Timber piles		
Isolated steel piles	Identify preliminary piles (Note P11)	State section reference or mass per metre and cross-sectional dimensions

3rd Division
The following separate items, each indicated by a bracketed letter below, are required for each group of piles
(a) Number of piles   nr   Give item for the number of piles in the group (Note P4)
(b) Depth driven   m   Measure driven depths along axes of the piles from the Commencing Surface to the bottoms of the toes of the piles (Note P6) and give item for the total depth driven for the group of piles (Note P4) Where piles are raked, state angle of rake within ranged increments of 10 degrees (Note P11).
(c) Length of piles   m   Measure length of piles expressly required (Note P7) and give item for the total length of piles in the group divided into the ranges of length given in the third division features of the CESMM (Note P4). Include the whole length of each pile in the item in the range of which the item occurs (Note P8).

COMMENTARY

Preformed Concrete, Timber and Isolated Steel Piles (Refer to Table 12.03)

The several separate items which require to be given for each group of preformed piles are noted in Table 12.03 which gives also the units of measurement. For other items in connection with preformed piles, see Class Q.

Items for depth driven give the total depth driven for each group of piles with no distinction as to depth ranges.

The measurement of the lengths of preformed piles affords an exception to the general rule that quantities are measured nett. The lengths measured are those expressly required. The items for the length of piles give the total of the expressly required pile lengths for each group of piles. Upon admeasurement, the lengths measured are those ordered whatever the finished lengths in the work may be. Items for cutting off surplus lengths are given in Class Q. The specification will usually reserve for the Employer the ownership of cut off lengths for use in any lengthening of piles that may be required. The ownership and disposal of cut off lengths surplus to lengthening requirements should be specified. Where the surplus is to become the property of the Contractor and is considered to have a credit value, special items may be included for the Contractor to offer rates for the credit value of the surplus.

COMMENTARY

Isolated Piles part in Tidal Water

Specimen item descriptions are given below for the isolated steel piles, part in tidal water, for the jetty shown in cross-section in the adjoining diagram. Piles assumed to be of the section and size stated in specimen items. Length of piles assumed for the purpose of specimen items.

MHWS

MLWS

Original Surface and Commencing Surface

(c)   (b)   (a)

Number	Item description	Unit
	PILES Isolated steel piles, steel to BS. 4360, grade 50, "H" section, 305 x 305 mm, mass 137 kg/m	
P751.1	Number of piles.	nr
P752.1	Depth driven, raked 10 - 20 degrees from vertical.	m
P755.1	Length of piles 10 - 15 m.	m
	PILES - Work affected by tidal water, high water level ordinary spring tides assumed 2.05 m above ordnance datum, low water level ordinary spring tides assumed 1.65 m below ordnance datum Isolated steel piles, steel to BS. 4360, grade 50, "H" section, 305 x 305 mm, mass 137 kg/m Driven at positions affected between high and low tide	
P751.2	Number of piles.	nr
P752.2	Depth driven.	m
P755.2	Length of piles 10 - 15 m.	m
	Driven at positions affected at all times	
P751.3	Number of piles.	nr
P752.3	Depth driven.	m
P755.3	Length of piles 10 - 15 m.	m

COMMENTARY

Where work is not otherwise described it is inferred that it is performed on land or above high water level.

The specimen items here given suffix Code number .1 are indicated (a) in the above diagram.

Item descriptions distinguish work which is affected by bodies of water. Item descriptions for work affected by tidal water distinguish between work affected at only some states of the tide and work affected at all times. Item descriptions state the water surface levels adopted for such distinctions. See Paragraph 5.20 of the CESMM.

Piles affected between high and low tides indicated (b) in above diagram. Those affected at all times indicated (c) in the diagram.

Any coating of piles required to be carried out before delivery to the Site would be made clear and included in the items for the piles. Any coating required to be carried out on Site would be measured under CESMM, Class V.

Fig. P1. Isolated steel piles for jetty part in tidal water.

Table 12.04 Interlocking Steel Piles

1st Division	
Interlocking steel piles	State make or type and the section reference.
2nd Division	State section modulus in cm3/m
3rd Division	
The following separate items, each indicated by a bracketed letter below, are required for each group of piles	
(a) Length of special piles, if any m	Give item for total length of each type of special piles (Note P5). State type of special piles in the item descriptions (Note P10). Measure length as that which is required (Note P7). Class corner, junction, closer and taper piles as special piles (Note P10). Measure closure and taper piles only when expressly required (Note P10).
(b) Driven area m2	Calculate area by multiplying the undeveloped length of the pile walls formed (including the length occupied by special piles) by the driven depth measured along the axes of the piles from the Commencing Surface to the bottom of the toes of the piles (Note P9).

(c) Area of piles of length: not exceeding
           5 m   m2
     5 - 10 m   m2
   10 - 15 m   m2
   15 - 20 m   m2
   20 - 25 m   m2
stated exceeding
     25 m   m2
	Calculate area by multiplying the undeveloped length of the pile walls formed (including the length occupied by special piles) by the lengths of the piles expressly required. (Note P9). Include the whole length of pile in the item in the range of which the length occurs (Note P8).

COMMENTARY

Interlocking Steel Piles (Refer to Table 12.04)

The several separate items given for each group of interlocking piles are noted in Table 12.04.

The piles themselves are measured by area in m2. The area is calculated as noted against (c) in Table 12.04. The length used to calculate area and for the length classification is that which is expressly required (another exception to the general rule that quantities are given nett). Items give the total area for each group of piles of particular type divided into items which distinguish the different Third Division length ranges.

Special piles are measured as part of the area of the piles. Additionally they are made the subject of separate items which give the total length of each type in metres. See (a) in Table 12.04.

Items for driven depths are measured as noted at (b) in Table 12.04.

COMMENTARY

Interlocking Steel Piles (cont.

     Where interlocking piles are used as the anchorage for anchored interlocking steel pile walls, they are given separately from the piles in the walls and are suitably identified.

     Steel walings and tie rods for anchored pile walls are excluded from Classes P and Q of the CESMM. They are classified by features and are each given as numbered items as provided in Class N. Item descriptions state or otherwise identify the grade of steel, the section reference, the cross-sectional dimensions, the mass per metre, the length and details of any off Site coating requirements. Any fish plates, spacers, back bolts and similar assembly fittings in connection with walings are identified and included in the numbered items of the walings. The bolts, washers and any couplings used for the tie tods are similarly included in the numbered items for the tie rods. (See Notes N1 and N2 of the CESMM).

PILING ANCILLARIES - CESMM CLASS Q

     The pile shafts and work installing them by boring or driving are covered by Class P of the CESMM. Work outside that covered by Class P which is incidental to piling is classified "piling ancillaries" and is catered for by Class Q. Note Q1 of the CESMM restricts the application of Class Q to work in the Class which is expressly required. Consequently, work covered by items in the Class which the Contractor chooses to carry out beyond the limit of that prescribed in the Contract or ordered by the Engineer is disregarded when work is admeasured.

Table 12.05 Piling Ancillaries

Generally	Measure work in this Class only where expressly required (Note Q1)		
1st Division	2nd Division		3rd Division
Cast in place concrete piles	Pre-boring	m	State diameter range as CESMM third division features
	Boring through rock	m	
	Backfilling empty bore with stated material	State material m	
	Permanent casing of stated thickness	State casing thickness m	
	Placing concrete by tremie pipe	m Measure full length of any pile where this method of concreting is used for the whole or part of the pile (Note Q2)	
	Enlarged bases	nr State diameter of enlarged bases and that of shaft from which enlarged (Note Q3)	
	Preparing heads	nr Preparing heads to receive pile extensions shall not be measured (Note Q7)	
	Reinforcement cages	t State the number and the diameter of the bars and the material of which composed (Note Q4)	

Table 12.05 Piling Ancillaries (cont.

1st Division	2nd Division		3rd Division
Preformed concrete piles	Preboring m		State cross-sectional area range as CESMM third division feature
	Jetting m		
Timber piles	Filling hollow piles with concrete m	Does not apply to piles comprising light gauge casings which are filled with concrete (Note Q5)	
	Number of pile extensions nr	Give separate items for the number and the length of extensions. State in the numbered item when piles are extended using material arising from cutting off surplus lengths of other piles. Measure driving extended piles under Class P. Do not measure length formed with surplus cut off material. Include length of scarved or other joints with length of timber pile extensions (Note Q7).	
	Length of pile extensions m		
	Cutting off surplus lengths nr		
	Preparing heads nr	Preparing heads to receive pile extensions shall not be measured	
Isolated steel piles			State mass range as CESMM third division features
Interlocking steel piles			State section modulus range as CESMM third division features
Delays and obstructions h	Standby driving rig: for driven cast in place piles for preformed concrete piles for timber piles for isolated steel piles for interlocking steel piles		
	Standby boring rig		
	Boring through artificial obstructions		
Surplus excavated material for disposal m3	Calculate volume from the nominal cross-sectional area of the piles and their length measured in accordance with Note P7. Include the volume of enlarged bases. Measure disposal of surplus excavated material only for cast in place piles (Note Q8)		

Table 12.05 Piling Ancillaries (cont.

1st Division	2nd Division		3rd Division
Pile tests nr	Kentiledge test of preliminary piles Anchor tests of preliminary piles Kentiledge tests of working piles Anchor tests of working piles	State particulars and rate of loading (Note Q9) State when load is not applied vertically (Note Q9)	State test load range as CESMM third division features

COMMENTARY

Piling Ancillaries (Refer to Table 12.05)

Subject to the work they represent being expressly required, items for piling ancillaries are measured under Class Q. Descriptive features, units of measurement and references to the Notes applicable to the work, as given in the CESMM, are noted or referred to in Table 12.05.

The measurement of an item in Class Q does not affect the conventions which govern the measurement of the piles and the boring and driving in Class P. For example, where a linear quantity of boring through rock is measured in Class Q, it is not deducted from that of the depth bored or driven given in Class P.

Placing Concrete by Tremie Pipe

Separate items from those for the piles are given for placing concrete by tremie pipe where this method of placing is expressly required. The length measured for the work is noted against the descriptive feature in the 2nd Division panel of Table 12.05. Placing concrete by tremie pipe is not measured in circumstances where the method is used because the Contractor of his own volition uses a drilling fluid to maintain the stability of the pile excavation.

Enlarged Bases

Lengths occupied by enlarged bases, where required, are measured and included in the lengths given in Class P for both the length of piles and the depth bored or driven. The numbered items given in Class Q for enlarged bases are intended to cover work, additional to that in the piles, involved in their construction. To make this clear, preamble is usually included in the Bill of Quantities stating items for enlarged bases are deemed to include the additional excavation, concrete and other work required for their construction. Item descriptions state the diameter of the base and that of the pile from which it is enlarged. Surplus excavated material arising from enlarged bases is measured as provided in Note Q8 of the CESMM. (See notes against the descriptive feature in Table 12.05).

Reinforcement

Reinforcement and prestressing steel is identified and included in the items for preformed concrete piles measured under Class P. Reinforcement cages for cast in place piles are given under Class Q. They are given by mass in tonnes in separate items from those for the piles. Details to be stated in the item descriptions are noted against the descriptive features in Table 12.05.

COMMENTARY

Piling Ancillaries (cont.

Pile Extensions

Separate items for the number of pile extensions and the length of pile extensions are given where the extensions are formed with material other than that cut off from other piles. Where the extensions are formed with material arising from that cut off from other piles, items for the number of pile extensions are given, they state in the descriptions that the extensions are so formed. Items for the length of piles are not measured where the extensions are formed with material cut off from other piles. The work of connecting the pile extensions to the piles to be extended is identified in the description of the items which give the number of pile extensions. (See notes against the descriptive feature for pile extensions in the 2nd Division panel of Table 12.05).

Cutting Off Surplus Lengths

Items give the number of surplus lengths to be cut off. The Specification will usually reserve for the Employer the ownership of cut off lengths for use in any lengthening of piles which may be required. The ownership and disposal of cut off lengths surplus to lengthening requirements should be specified. Where the surplus is to become the property of the Contractor and is considered to have a credit value special items may be included for the Contractor to offer rates for the credit value of the surplus.

Delays and Obstructions

An estimate of the total of the periods in units of hours during which it is likely that rigs will standby or it is likely will be spent in dealing with obstructions consequent upon the instructions of the Engineer, is attached to the items given in the Bill of Quantities.

Upon admeasurement, the periods which qualify to be measured are those which result from the instructions of the Engineer. The items for the standing by of rigs is appropriate where this occurs due to the work being suspended by an instruction of the Engineer. The items for boring through artificial obstructions is considered to include any consequential standing by of rigs during the period of dealing with obstructions. This is made clear in preamble.

Surplus Excavated Material for Disposal

The measurement of disposal of surplus excavated material from piling operations is restricted to that which arises from the installation of cast in place piles. The volume in m3 for disposal is calculated as provided in Note Q8 of the CESMM. (See notes against the descriptive feature in the 2nd Division panel of Table 12.05). Disposal requirements should be given in the specification. Where such requirements are for disposal on Site, the location of the disposal area is stated in the item descriptions.

Pile Tests

It is usual to relate items for pile tests to specification clauses wherein the tests and other requirements are described. The Bill should make clear that the items for tests are deemed to include all requirements specified for the tests including preparing piles for testing.

EXAMPLE No PE.1.

Measured Example

A measured example of bored cast in place piles follows.

172

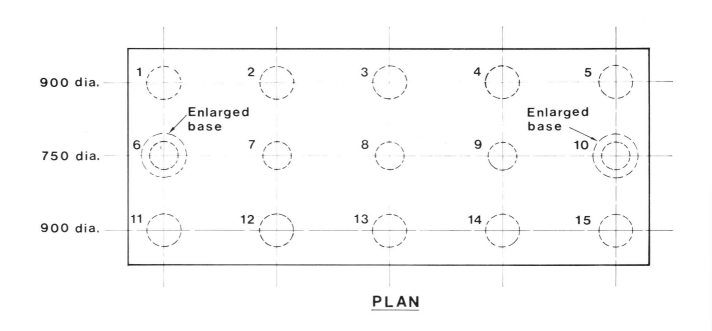

PLAN

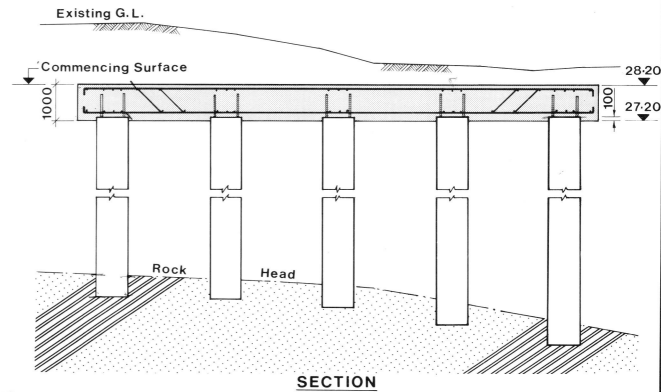

SECTION

Pile No.	1	6	11	2	7	12	3	8	13	4	9	14	5	10	15
Rock Level	19·60			19·50			19·30			18·80			18·30		
Base Level	19·00			18·90			18·70			18·20			17·70		

NOTES

1. Specified cut-off level 27·30
2. Concrete to be Grade 25, cement to B.S. 4027

PILE FDN. DRG. No. P/D/1				
EXAMPLE PE.1				1000 / 100 / 900

	28.20	28.20	28.20	28.20	28.20
	19.00	18.90	18.70	18.20	17.70
D.Bored	9.20	9.30	9.50	10.00	10.50
	0.90	0.90	0.90	0.90	0.90
Length	8.30	8.40	8.60	9.10	9.60

Piles
Bored cast in place
conc. piles, conc. grad.
25 ct. to BS.4027

Diam. 750 mm

5/ 1   Number of piles (P141

{ Length of piles (P142

8.30
8.40
8.60
9.10
9.60

{ Depth bored 5-10 m;
{ Comm. Surf. top of
{ pile cap base (P144.1

9.20
9.30
9.50
10.00

10.50   Depth bored 10-15 m;
Comm. Surf. top of
pile cap base (P145.1

Diam. 900 mm

5/ 2   Number of piles (P151

{ Length of piles (P152

2/ 8.30
2/ 8.40
2/ 8.60
2/ 9.10
2/ 9.60

(1)

COMMENTARY

Rather than make waste calculations on the dimension sheets as here shown, it will be found convenient for the more extensive installations to prepare a separate schedule indicating for each pile the length, depth bored or driven, work required to heads, bases and the like, in a manner which will allow of like items being collected together for transfer to dimension sheets or for billing direct.

The precise diameters of the piles are stated in item descriptions, the third division ranges are over-ridden by Note P1 of the CESMM.

The items assume that the site is to be reduced to the level of the top of the pile caps prior to piling commencing. This will form the working surface and it will also be the Commencing Surface for boring. The Commencing Surface is not the Original Surface and is, therefore, identified in the item descriptions.

The length of piles measured is the concreted length measured from the bottom of the shafts to the specified cut off levels. It includes the length occupied by enlarged bases but excludes any cut off tolerances. It is assumed that there is no specified requirement for the piles to be concreted to a level higher than the specified cut off levels for subsequent removal. Numbered items for each type of enlarged base are required notwithstanding that the length occupied by the bases is included in the pile length. These items are given subsequently. The length occupied by any cut off tolerance is neglected when measuring the lengths of piles.

The depth range for boring given in the item descriptions is that in which the whole depth of boring for each pile occurs.

PILE FDN – DRG. No. P/D/1

Bored piles (cont.

Diam. 900 mm (cont.

2/ 9.20	Depth bored 5 – 10 m
2/ 9.30	Comm. Surf. top of
2/ 9.50	pile cap base (P154.1
2/ 10.00	
2/ 10.50	Depth bored 10 – 15 m
	Comm. Surf. top of
	pile cap base (P155.+

Piling Ancills.

```
              9.20
              9.30
   2 ) 750    9.50
       375   10.00
   2 ) 900   10.50
       450   48.50
```

22/7/ 0.38	Surplus excavd. matl.
0.38	for disposal (Q700
48.50	
2/ 22/7/ 0.45	
0.45	
48.50	

Rock
Tos

19.60	19.50	19.30	18.80	18.30
19.00	18.90	18.70	18.20	17.70
0.60	0.60	0.60	0.60	0.60

Cast in place conc. piles

5/ 0.60	Boring thro. rock, diam. 750 mm. (Q124
2/5/ 0.60	Boring thro. rock, diam. 900 mm (Q125

(2)

COMMENTARY

The intention of the single line drawn across the dimension column of the preceding column of dimensions is to indicate the end of the items to which the second sub heading applies (i.e. Diam.) and to leave the first sub heading (Bored cast in place piles) operative for subsequent diameters. When the end of the items for the several diameters is reached a double line is drawn across the dimension column to indicate the end of the items to which both sub headings apply.

The dimensions taken for the volume of surplus excavated material for disposal are the nominal diameter of the piles and the depth bored. The additional volume created by enlarged bases would be added to Item Q700. The additional volume is measured following the items for enlarged bases which are taken later in this Example. Disposal means disposal off site unless otherwise stated.

A definition of rock would be given in the preambles to the Bill of Quantities.

PILE FDN. DRG. NO. P/D/1

Cast in place conc. piles (cont

piles 9+10

	9.10	{ Placg. conc. by tremie
	9.60	{ pipe, diam. 750 mm (Q154

piles 4,5,14+15

2/	9.10	{ Placg. conc. by tremie
2/	9.60	{ pipe diam 900 mm (Q155

bases.

piles 6+10

2/   1   Enlarged bases, diam. 1200 mm to shafts diam. 750 mm   (Q164

$$\frac{1200-750}{2} = 225 \times \text{Cot. } 60°$$

$225 \times 1.732 = \underline{390} \text{ hght}$

$1200 + 750 = 1950 \div 2 = 975$

$975 \div 2 = \underline{488} \text{ av. radius}$

2/22/7	0.49
	0.49
	0.39

Surplus excavd. matl. for disposal   (Q700

2/22/7	0.38
	0.38
	0.39

Ddt. Last item (Q700

(shaft.

Rfcmt.

$1000 \div 150 = 6.666$

$750 - (2 \times 40) = 0.670$

$0.670 \times \pi = 2.106$

$2.106 \times 6.666 = 14.04$

$14.04^2 + 1.00^2 = 198.12$

$\sqrt{198.12} = 14.08$

Helical 1 m lgth 750 ϕ = 14.08

$900 - (2 \times 40) = 820$

$0.820 \times \pi = 2.577$

$2.577 \times 6.666 = 17.178$

$17.18^2 + 1.00^2 = 296.15$

$\sqrt{296.15} = 17.24$

Helical 1 m lgth 900 ϕ = 17.21

COMMENTARY

It is assumed that water will be encountered in piles Nos, 4,5,9, 10,14 and 15 and that the specified requirement is to place the concrete by tremie pipe in these situations. Placing concrete by tremie pipe is measured for the whole length of a pile irrespective of whether it is used for the whole or part of the pile.

Enlarged bases are required to piles Nos. 6 and 10. Items for enlarged bases state the diameter of the base and the diameter of the shaft from which they are enlarged.

It is assumed that the specified requirement is for the frustum of the enlargement to make an angle of 60 degrees to the horizontal as shown on the sketch below. The waste calculations preceding Item Q700 calculate by Trigonometry the height of the enlarged bases.

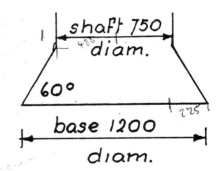

The reinforcement details, (See next column) show the reinforcement cages to be of vertical bars and helical bindings. The wastes under the heading "Rfcmt" calculate the developed length of the helical spiral for one metre length of pile of each diameter. The workings in the wastes calculate a single circumference and then multiply it by the number there would be at the 150 mm pitch in a one metre length of pile. The total of the circumference so obtained represents the base of a triangle, the height (one metre) is taken as the perpendicular and the hypotenuse (found by the Theorem of Pythagoras) is the developed length of the spiral.

PILE FDN. - DRG. No P/D/1

Cast in place conc. piles (cont.

	Rfcmt. per m. 750 ⌀	Kg
	3.853 × 16	61.65
	0.616 × 14.08	8.67
	Kg. per m	70.32
	Rfcmt per m 900 ⌀	Kg
	6.313 × 16	101.01
	0.616 × 17.21	10.60
		111.61

	8.30	
	8.40	
	8.60	
	9.10	
	9.60	
5/	0.50	

Rfcmt. cages; 16 No 25 mm diam. HY steel bars + 10 mm diam MS helical binds. (Q180.1

(proj. top

× 70.32 =   Kg

2/	8.30	
2/	8.40	
2/	8.60	
2/	9.10	
2/	9.60	
2/5/	0.50	

Rfcmt cages; 16 No 32 mm diam. HY steel bars + 10 mm diam MS helical binds (Q180.2

(proj. top

× 111.61 =   Kg

Heads

5/	1

Prepare heads, diam. 750 mm   (Q174

2/5/	1

Prepare heads, diam. 900 mm   (Q175

COMMENTARY

Details of the reinforcement cages taken in the Example are illustrated in the following diagram

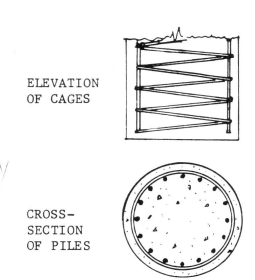

ELEVATION OF CAGES

CROSS-SECTION OF PILES

The cages for the 750 mm diameter piles comprise 16 No 25 mm diameter vertical bars with 10 mm diameter helical binding at 150 mm pitch. The cages for the 900 mm diameter piles are similar except that the vertical bars are 32 mm diameter. The waste calculations which precedes the dimensions for the cages, calculates the mass of reinforcement per linear metre of each size of pile. The dimensions set down for the cages are the length of the piles in linear metres. Provision is made for total length to be multiplied by the mass per metre to calculate the total mass in kg. of each type of cage. The mass would be reduced to tonnes when billing the cages.

Items for preparing heads are here measured on the assumption that the Specification calls for some work in connecting the heads to the pile cap. In cases where no work is expressly required to the pile heads the items would not be measured.

PILE FDN. - DRG. No. P/D/1

Delays + obstructns.

h	
10	Standing by of boring rig; per rig (Q660
h	
10	Boring thro. artificial obstructns. (Q670

Pile tests

1	Anchorage test; ML proof loading test on working pile as Spec. clause "X", test load 200-300t (Q843

COMMENTARY

For commentary on "Delays and obstructions" and "Pile tests" refer to Commentary under the heading "Piling Ancillaries" in this Chapter.

Work to the pile cap is not measured in the Example. Excavation for the cap would be classified "Excavation of foundations", stating it was around pile shafts (See Table 7.01 Chapter 7). A deduction from the volume of this excavation would be made for the volume of boring, between the top and base of the cap, measured with the piles.

Concrete in the pile cap and related ancillaries would be measured as provided in Classes F and G of the CESMM, respectively.

13 Roads and Pavings—CESMM Class: R

Briefly summarised, the work covered by Class R includes sub-base, base and surfacing of roads, runways and other paved areas, together with kerbing, light duty pavements, footways, cycle tracks, traffic signs and surface markings. Earthworks to receive the work and any landscaping are measured in accordance with Class E. (See Chapter 7). Drainage for roads and the like is measured in accordance with Classes I, J, K and L (See Chapter 9). Fencing is measured in accordance with Class X (See Chapter 17). Supporting posts and foundations for traffic signs are included in the items for traffic signs in Class R. Gantries and other substantial structures to support traffic signs are measured in detail in accordance with the Classes appropriate to the particular work in them.

The classification table for Class R, includes descriptive features for bases, surfacing and concrete pavements. The features for heavy duty work of this description for roads, aircraft runways and such like, are those given in the First Division of CESMM Class R, Codes 1 - 4 and their attached Second and Third Division features. Separate descriptive features applicable to work of this description for light duty pavements, footways, cycle tracks and the like, are given in the First Division of CESMM Class R, Code 7 and its attached Second and Third Division features.

For work included in Class R and for exclusions from the Class refer to the "Includes" and "Excludes" given at the head of the Class R classification table in the CESMM. References to work included in Class R are made in the "Excludes" for other Classes at the head of the classification tables in the Classes F, G, H and N of the CESMM.

Table 13.01 Class R

Generally The expression "DoE Specified" means as specified in the "Specification for Road and Bridge Works" issued by the Department of the Environment (Note R13). See subsequent Commentary.

COMMENTARY

D.o.E. Specified

The current edition of the Specification for Road and Bridge Works is issued by the Department of Transport. Where this is to be used, Preamble amending the CESMM needs to be included in the Bill of Quantities. Suggested preamble follows:-

For the purpose of this Bill of Quantities, Note R13 of the CESMM shall be deemed to be deleted and the following new Note R13 shall be deemed to be inserted in its place. "The expression DoE specified means as specified in the Specification for Road and Bridge Works, Fifth Edition, 1976, as extended and amended by Supplement No. 1, 1978, issued by the Department of Transport and obtainable from Her Majesty's Stationery Office".

COMMENTARY

D.o.E. Specified (cont.

 Unless otherwise stated, the statement "DoE Specified" in any item description has the effect of incorporating in the Contract those requirements prescribed
in the "Specification for Road and Bridge Works" (issued by the Department of
the Environment) which relates to the material covered by the item.

Table 13.02 Sub-bases, Flexible Road Bases and Surfacing

1st Division	2nd Division		3rd Division
Sub-bases, flexible road bases and surfacing	Classify in accordance with 2nd Division features m2	Identify materials for all courses (Note R1) Measure width of each course of material at the top surface of the course (Note R3) A manhole cover or other intrusion less than one square metre in area shall not be deducted (Note R3) State when applied to surfaces inclined at an angle exceeding 10 degrees to the horizontal (Note R2)	State actual depth of each course (Note R1)
	Additional depth m3	State nature of material	
	Regulating course t	State nature of material	

COMMENTARY

Sub-bases, Flexible Road Bases and Surfacing (Refer to Table 13.02)

Each course of material is measured superficially and given in m2. The material
in each course is identified and the actual depth or thickness of the course is
stated in the item description. Stating the actual depth or thickness is a
requirement of Note R1 which overrides the Third Division features of Class R
which give depth ranges. Layers, in courses of the same material specified to
be spread and compacted in layers, are not classed as courses.

 Where different tolerances in surface levels or different consolidation
requirements, etc., are prescribed for a course of the same material, items with
additional description distinguish each difference.

COMMENTARY

Sub-bases, Flexible Road Bases and Surfacing (cont.

Items for additional depth of stated material quantified by volume in
m3 are given where additional depth is required in intermittent or isolated
areas below the uniform depth of a course.

Items are given for regulating courses where they are required. Materials
are stated in the item descriptions and quantities are given by mass in tonnes.

Concrete Pavements (Refer to Table 13.03)

The descriptive features given in the CESMM for concrete pavements generate
separate items for concrete slabs, for any reinforcement in the slabs and
for any waterproof membrane below the pavements. See Table 13.03 which notes
or refers to the descriptive features, the units of measurement and the details
which require to be given in item descriptions.

Quantities for the items for concrete slabs give the area in m2 for each
depth of slab. The depth being the actual depth, as required by Note R1. Where
the specification calls for a concrete slab of monolithic construction to be
spread and compacted in layers and/or for the slab to combine aerated and non-
aerated concrete, whilst the make up of the slab must be identified, it is not
necessary to separately itemise each layer. The depth given for a particular
slab of this nature is the aggregate of the depths of its layers.

The absence of any provision in Class R for the separate itemisation of
the finish to the upper surface of concrete carriageway slabs infers that the
work of finishing the top surface of the concrete is intended to be part and
parcel of the items for the carriageway slabs. Added description or item
coverage preamble will usually make this clear.

The materials for waterproof membranes are identified in the item
descriptions. Membranes are given in m2. In accordance with the general
philosophy of the CESMM they are measured nett with no allowance for the
additional material in laps. Item descriptions or Preamble state they have
been measured as here stated.

Joints in Concrete Pavements

Detail drawings are invariably provided for joints in concrete pavements. To
comply with Note R7 of the CESMM, which requires certain information to be given
(See 3rd Division, Table 13.03), it is considered preferable for item descriptions
to make reference to drawings from which the information can be obtained rather
than to detail it in lengthy descriptions.

Measured items, quantified in linear metres, are given for all joints in
concrete pavements, other than construction joints which are not expressly
required. (See last sentence of Note R7 of the CESMM). The Second Division
features use the same terms as the "Specification for Road and Bridge Works" to
describe the joints. Formwork to joints is not separately itemised (See "Generally",
Table 13.03). Item descriptions are required to state the depth range of the
joints as set out in the 3rd Division of the classification table. The person
preparing the Bill may choose to state actual depth of joints in preference to
depth range, where he considers it would be more helpful to do so. Actual depth
is stated for joints of a depth exceeding 300 mm and also where there is but one
depth in one item.

Table 13.03 Concrete Pavements and Joints in Concrete Pavements

Generally Separate items are not required for formwork to pavements or formwork to joints in pavements (Note R6)

1st Division	2nd Division			3rd Division
Concrete pavements	Carriageway slabs of DoE Specified paving quality concrete m2	State strength of concrete (Note R1)	No deduction for a manhole cover or other intrusion less than one sq.m. in area (Note R3)	State actual depth of slab (Note R1) State where inclined at an angle exceeding 10 degrees to the horizontal (Note R2)
	Other carriageway slabs of stated strength m2			
	Other in situ concrete slabs of stated strength m2			
	Steel fabric reinforcement to BS 4483 m2	State type number or letter in accordance with BS 4483 (Note R4) Measure nett with no allowance for laps (Note R4)		Classify according to mass per sq.m. as 3rd Division features
	Other fabric reinforcement m2	State the material (Note R4) Measure nett with no allowance for laps (Note R4)		State nominal mass per sq.m.
	Steel bar reinforcement to BS 4449 or BS 4461 t			State diameter as 3rd Division features. Classify bars not circular in cross-section by the diameter of the circular bar in classification nearest in cross-section to the bars measured (Note R5)
	Other bar reinforcement t	State material		
	Waterproof membrane below concrete pavements m2	State material		State where inclined as Note R2
Joints in concrete pavements	Longitudinal m			State depth of joint State the dimensions, spacing and nature of components (Note R7)
	Expansion m			
	Contraction m			
	Warping m			
	Butt m			
	Construction m	Measure only where expressly required (Note R7)		

COMMENTARY

Concrete Pavements (cont.

Work Applied to Inclined Surfaces

The rule stated in Note R2 of the CESMM requires that item descriptions for work in Class R which is "applied to surfaces inclined at an angle exceeding 10 degrees to the horizontal" shall so state. The rule is considered not applicable to reinforcement in concrete slabs, joints in concrete pavements and traffic signs, because it is thought inappropriate to describe this work as "applied to surfaces".

Intrusions

The rule in the second sentence of Note R3 of the CESMM provides that "The area of a manhole cover or other intrusion into a surface shall not be deducted where the area of the intrusion is less than one square metre". As area could only be deducted from area and the rule refers to a surface, it can be said that the rule applies only to an item of work which is measured by area and which creates a surface rather than surfaces. If this is accepted it makes the rule inapplicable to reinforcement in concrete slabs.

Table 13.04   Kerbs, Channels and Edgings

1st Division	2nd Division	3rd Division	
Kerbs, channels and edgings	Classify in accordance with 2nd Division features State materials, cross-sectional dimensions and details of backings and beds (Note R8) State where inclined at an angle exceeding 10 degrees to the horizontal (Note R2)	Straight or curved to radius exceeding 12 m	m
		Curved to radius not exceeding 12 m	m
		Quadrants	nr
		Drops	nr
		Transitions	nr

COMMENTARY

Kerbs, Channels and Edgings (Refer to Table 13.04)

Item descriptions other than those which identify requirements by B.S reference, state or otherwise identify materials, cross-sectional dimensions and type of kerb, channel or edging. Item descriptions also state materials, cross-sectional dimensions and details of the required beds and backings.

Kerbs, channels and edgings are measured in linear metres. A single classification combines curved work of a radius exceeding 12 m with that which is straight. Work curved to a radius not exceeding 12 m is given separately. Additional enumerated items are given for quadrants, drops (each dropper is measured) and for transitions. The standard features for these enumerated items are amplified to identify requirements where it would not otherwise be clear. Related earthworks are excluded from the items. Minor earthworks, such as a trench, may be included by a specific statement to that effect in the description and a note in Preamble amending the CESMM.

Table 13.05 Light Duty Pavements

1st Division	2nd and 3rd Divisions
Light duty pavements	Refer to Class R, Code 7, 2nd and 3rd Division features The notes in Tables 13.02 and 13.03 are applicable to light duty pavements

Table 13.06 Traffic Signs and Surface Markings

1st Division		2nd Division		
Traffic signs and surface markings	State materials, size and diagram number taken from traffic signs regulations and general directions issued by DoE. (Note R10)	Traffic signs:- Non-illuminated nr Illuminated nr	Separate items are not required for supporting posts and foundations to traffic signs. Measure other substantial structures associated with traffic signs as provided in other appropriate Classes (Note R11)	
		Surface markings:- Non-reflecting road studs nr		State where applied to surfaces inclined at an angle exceeding 10 degrees to the horizontal (Note R2)
		Reflecting road studs nr	State shape and colour aspects (Note R9)	
		Letters and shapes nr Continuous lines m		
		Intermittent lines m	Exclude gaps from lengths measured (Note R12)	

COMMENTARY

Traffic Signs and Surface Markings (Refer to Table 13)

The descriptive features for traffic signs and surface markings and the units of measurement are set out in Table 13.06. The materials, size and diagram number (taken from *Traffic signs, regulations and general directions, issued by the Department of the Environment) are added to the features in the item descriptions.

Separate items are not required for supporting posts and foundations for traffic signs. Item descriptions identify the posts and foundations. Otherwise similar traffic signs with different posts and foundations are given in separate items which distinguish each difference. Item descriptions for traffic signs supported on gantries or structures which have been measured separately from the signs, state that the gantries or structures are measured separately.

Where similar surface markings are applied to different base surfaces which involve differences in application and where different preparation is specified for the same base surface, item descriptions distinguish differences by stating or otherwise identifying the base surface or the preparation.

Road studs are classified "non-reflecting" or "reflecting". Item descriptions for reflecting road studs state the shape and colour aspects of the studs.

Surface marking with lines is given in linear metres. Item descriptions state the width of line. The length measured excludes the gaps in intermittent line markings.

EXAMPLE RE.1

Specimen Bill Items

The Example which follows gives specimen items from a Bill of Quantities for a motorway. Units of measurement are given against the items. The quantities have been omitted.

* *Now superseded by Statutory Instrument 859.*
  See pages 3 and 4 of Example RE.1

Number	Item description	Unit	Quantity	Rate	Amount	
					£	p
	ROADS AND PAVINGS					
	Sub-bases, flexible road bases and surfacing, DoE Specified Sections 800 and 900			COMMENTARY		
R118	Granular material type 1, DoE Specified clause 803, depth 450 mm; sub-base.	m2		The specimen Bill items in the Example have been drafted on the		
R124	Granular material type 2, DoE Specified clause 804, depth 150 mm; sub-base.	m2		assumption that the Particular Specification provides that the Specifica-		
R155	Lean concrete, DoE Specified clause 807, depth 200 mm; to roadbase to carriageways.	m2		tion for the work shall be the "Specification for Road and Bridge Works"		
R225	Dry bound macadam, DoE Specified clause 809, depth 200 mm; roadbase to hardshoulders.	m2		(See earlier Commentary) and provides also for additional clauses for work not covered by the quoted Specification.		
R232.1	Dense bitumen macadam, DoE Specified clause 903, depth 60 mm; basecourse to carriageways.	m2		Although not strictly necessary if the descriptions otherwise clearly		
R232.2	Dense bitumen macadam, DoE Specified clause 903, depth 60 mm; basecourse to hardshoulders.	m2		identify the work they represent, it is considered helpful, as shown in the Example, to make		
R322.1	Rolled asphalt, DoE Specified clause 907, depth 35 mm; wearing course to hardshoulders.	m2		reference to the appropriate Sections and clauses of the Specification.		
R322.2	Rolled asphalt, DoE Specified clause 907, depth 40 mm; wearing course to carriageways.	m2		Items are given for each course of base or surfacing. They identify the		
R341.1	Surface dressing with 20 mm nominal coated chippings, DoE Specified clause 907, depth 25 mm, on rolled asphalt surfaces of carriageways.	m2		materials and state the thickness in accordance with Note R1. To aid identification, added description has been given locating the courses.		
	Concrete pavements, DoE Specified Section 1000					
R417	Carriageway slabs of concrete, DoE Specified clause 1004, depth 280 mm, including surface finish DoE Specified clause 1021 and curing DoE Specified clause 1022.	m2				
	(1) To Part 2.1 Summary Page total					

Number	Item description	Unit	Quantity	Rate	Amount	
					£	p
	ROADS AND PAVINGS (cont.					
	Concrete pavements, DoE Specified Section 1000 (cont.				COMMENTARY	
R443	Steel fabric reinforcement to BS 4483, reference A193, nominal mass 3.02 kg/m2.	m2			Fabric reinforcement is measured nett with no allowance for laps. See Table 13.03.	
R480	Waterproof membrane below concrete pavements of 500 grade polythene film as Specification clause 10.11 (measured nett with no allowance for laps).	m2			Item descriptions for joints in concrete pavements are required to state the depth of joint, and the dimensions, spacing and nature of components. See Table 13.03. The specimen item descriptions in the Example use Specification and Drawing references from which information not stated in the description can be obtained.	
	Joints in concrete pavements, including reinforcement, dowel bars, grooves and sealing shown on the Drawings or specified					
R517	Longitudinal joints, DoE Specified clause 1010, depth 280 mm, as Drawing No. S/10/22.	m				
R527	Expansion joints, DoE Specified clause 1009, depth 230 mm, as Drawing No. S/10/23.	m			For the classification "Kerbs, channels and edging" the classification table for Class R in the CESMM provides 2nd Division descriptive features for precast concrete kerbs, edgings and channels, for in-situ concrete kerbs and edgings and for asphalt kerbs and channels. The specimen item descriptions state materials, cross-sectional dimensions and details of beds and backings as required by Note R8 of the CESMM. See Table 13.04.	
R537	Contraction joints DoE Specified clause 1009, depth 280 mm, as Drawing No. S/10/24.	m				
	Kerbs, channels and edging as Specification clause 11.9, with concrete grade 10/38 foundations and haunching including formwork to edges and haunching					
R651	Precast concrete channels to BS 340, size 150 x 125 mm, on 300 x 150 mm foundation and with 100 x 150 mm (extreme) haunching, straight or curved to radius exceeding 12 m.	m				
R652	Precast concrete channels to BS 340, size 150 x 125 mm, on 300 x 150 mm foundation and with 100 x 150 mm (extreme) haunching, curved to radius not exceeding 12 m.	m				
	(2) To Part 2.1 Summary Page total					

PART 2.1 MAIN CARRIAGEWAY

Number	Item Description	Unit	Quantity	Rate	Amount	
					£	p
	ROADS AND PAVINGS (cont.					
	Traffic signs and surface markings conforming to the diagrams or references stated in the item descriptions, which are taken from "The Traffic Signs, Regulations and General Directions 1981 (Statutory Instrument 859, obtainable from Her Majesty's Stationery Office)					
	Non-illuminated traffic signs as BS. 873: Part 2.					
R810.1	Marker posts, exposed height 975 mm, with reflective markers both sides, as Detail "B", Drawing No. S/12/2	nr				
	Non-illuminated traffic signs, manufactured from steel, vitreous enamel finish and with galvanised steel posts and fittings and mass concrete foundations all as BS. 873: Part 1, DoE Specified Section 1200					
R810.2	Reflectorised sign, size 2000 x 1125 mm, diagram 920, as Detail "C", Drawing No. S/12/2.	nr				
R810.3	Reflectorised sign, size 4500 x 1725 mm, diagram 917, as Detail "E", Drawing No. S/12/2.	nr				
	Externally illuminated traffic signs manufactured from steel, vitreous enamel finish and with galvanised steel posts and fittings, external lighting lanterns and wiring from point of supply and mass concrete foundations all as BS. 873: Part 1, DoE Specified Section 1200					
R820.1	Signs, size 1500 x 1500 mm, diagram 905, as Detail "F". Drawing No. S/12/3.	nr				
	(3) To Part 2.1 Summary Page total					

COMMENTARY

Materials for traffic signs and sizes need to be stated in item descriptions in addition to the diagram numbers from "The Traffic Signs, Regulations and General Directions 1981".

Descriptions as given in the Example may be shortened by reducing detail and giving Specification and Drawing reference, provided they clearly identify the work. Descriptions or Preamble should make clear the intended item coverage.

Number	Item Description	Unit	Quantity	Rate	Amount	
					£	p
	ROADS AND PAVINGS (cont.					
	Traffic signs and surface markings conforming to the diagrams of references, stated in the item descriptions, which are taken from "The Traffic Signs, Regulations and General Directions 1981"(Statutory Instrument 859, obtainable from Her Majesty's Stationery Office)					
	Road studs, to BS. 873:Part 4, as Specification clause 1204					
R840.1	"Catseye" white reflecting road studs, size 265 mm long, to concrete surfaces.	nr				
R840.2	"Catseye" red reflecting road studs, size 265 mm long, to asphalt surfaces.	nr				
R840.3	"Catseye" red reflecting road studs, size 265 mm long, to concrete surfaces.	nr				
	Surface markings of reflectorised thermoplastic material as BS. 3262: Part 1 DoE Specified Section 1200					
R850	Letters, height 1.6 m, reference Part X of Schedule 7, on asphalt surfaces.	nr				
R860.1	Longitudinal continuous lines, width 200 mm diagram 1012.1 on concrete surfaces.	m				
R860.2	Longitudinal continuous lines as marginal strip, width 200 mm, diagram 1012.1 on rolled asphalt surfaces.	m				
R860.3	Transverse continuous lines, width 200 mm, as diagram 1001, on asphalt surfaces.	m				
R870.1	Longitudinal intermittent lines, width 100 mm, 1000 mm lines and 5000 mm gaps, as diagram 1005, on concrete surfaces.	m				
	(4) To Part 2.1 Summary			Page total		

COMMENTARY

The surface to which road studs and surface markings are applied has an affect on cost. The specimen item descriptions distinguish the different surfaces to which the work is applied.

14 Rail Track—CESMM Class: S

Earthworks in connection with railway track work are measured under Class E. Together with work in other Classes associated with rail track, it is usually grouped in the same Part of the Bill of Quantities as the work measured under Class S. The classification "Filling and compaction to stated depth or thickness" in Class E is appropriate for blinding and blanketing of sand, ashes, gravel or similar material, of uniform thickness, where this is required to receive the ballast or other track foundation. Items for the preparation of the earthwork surfaces are not measured where blinding and blanketing which have been measured under Class E are in direct contact with the earthwork formation surface. (See Note E27 of the CESMM). Items for preparation of earthwork surfaces are measured as provided in Class E, where ballast and track foundations measured under Class S or materials other than that measured under Class E are in direct contact with the earthwork formation surfaces.

The work for which Class S is appropriate includes rails, chairs, sleepers, switches, crossings and track accessories as well as ballast and concrete track foundations. Overhead crane rails are specifically excluded from Class S and are included in Class M.

Table 14.01 Rail Track Generally: Class S

Generally: In Class S:-

"Supply" includes delivery of components to the Site (Note S1)

"Laying" comprises all work subsequent to delivery of components to the Site (Note S1)

Include details of materials to be supplied or laid (Note S3)

Track is deemed to be standard gauge (1435 mm) unless otherwise stated (Note S2)

Separate items are required for the supply of materials and assemblies and for the laying of track, where track is supplied and laid by the Contractor (Note S9)

Items for laying track are deemed to be for sleepered track laid on loose ballast unless otherwise stated (Note S11)

COMMENTARY

Rail Track Generally (Refer to Table 14.01)

The CESMM Notes of general application to Class S are referred to in Table 14.01.

Items for laying rail track and track assemblies are given separately from those for supplying the materials. Items for ballast and concrete foundation slabs include supplying and laying the materials for the work they describe.

Item descriptions for supply give full details of the materials and components to be supplied. See subsequent Commentary.

Item descriptions for laying only track, etc., identify type and general requirements without listing in detail the materials and components involved. Where materials and components are supplied to the Contractor by the Employer, descriptions make clear where the Contractor is to provide

COMMENTARY

Rail Track Generally (cont.

incidental materials or components, not supplied by the Employer, but which are intended to be included in the laying only items. The location where supplied components will be made available to the Contractor and his responsibility for taking over, unloading, transporting, assembly, etc., need to be stated. The foregoing details may be incorporated in a sub-heading in the Bill of Quantities or they may be covered by Specification clause or item coverage Preamble to which item descriptions or sub-headings refer.

Table 14.02   Ballast and Track Foundations

1st Division	2nd Division			3rd Division
Ballast and track foundations	Crushed stone	m3		
	Slag	m3		
	Gravel	m3		
	Other loose ballast	m3		
	Concrete track foundation slabs	m	State dimensions, specification of concrete, reinforcement and particulars of joints (Note S8)	
	Over width concrete track foundation slabs	m2	Measure length as that of the track laid on them (Note S8)	
			Separate items are not required for formwork and reinforcement or grouting up track (Note S10)	
	Bitumen filler to complete rail and angle assemblies	m		

COMMENTARY

Ballast and Track Foundations (Refer to Tables 14.01 and 14.02)

Ballast

Items for ballast give the quantities in m3. Materials used for ballast are stated or identified in the item descriptions. Placing and compacting require-ments are also identified in the descriptions. The differences between bottom ballast (that placed before the track is laid) and top ballast (that placed after the track is laid) make it advisable to implement Paragraph 5.10 of the CESMM and give each as separate items. Preamble or item descriptions should indicate whether or not the volumes of sleepers have been deducted from the volumes measured for the top ballast. The classification table for Class S of the CESMM makes no provision for the separate itemisation of forming the ballast to slopes at the sides or shoulders. It is taken that this work is intended to be included in the cube items given for the ballast. This is made clear by Preamble.

Concrete Track Foundation Slabs

Concrete track foundation slabs of uniform width are given in the Bill of Quantities in linear metres stating the width and the thickness. Over width foundation slabs are given in m2 stating the thickness. Item descriptions for concrete track foundation slabs state the specification of the concrete

COMMENTARY

Ballast and Track Foundations (cont.

Concrete Track Foundation Slabs (cont.

and the reinforcement. They state also particulars of the joints in the slabs.
Note S10 of the CESMM provides that separate items are not required for formwork and reinforcement for concrete track foundation slabs. No such provision
is made for joints and concrete accessories. In the absence of their specific
exclusion from Class G and in the absence of a Note in Class S to the effect
that separate items are not required for them, it is considered appropriate to
measure items as provided in Class G for any joints, for any inserts and for
any finishing to the top surface of concrete track foundation slabs.   The
finishing of top surface, carried out as part of the concrete placing process
and not given a separate finishing treatment is included in the items for the
concrete track foundation slabs by specific statement to that effect in the
item descriptions or in Preamble.

Table 14.03 Supply Only: Rails and Other Materials for Plain Track

1st Division	2nd Division			3rd Division
Supply only: Rails for plain track	Bullhead rails	t	State section reference given in BS 9* (Note S4)	Classify in accordance with 3rd Division range
	Flat bottom rails	t	State section reference given in BS 11(Note S4)	
	Dock rails	t	State section reference given in BS 2 (Note S4)	
	Guard angles	t	State dimensions and mass per metre (Note S4)	Classify in accordance with 3rd Division range
Supply only: Other materials for plain track	Timber sleepers	nr		
	Concrete sleepers	nr		
	Other sleepers	nr		
	Plain chairs	nr	State particulars	
	Other chairs	nr	of bolts, spikes, rods, keys, insul-	
	Bearing plates	nr	ators and other	
	Spacer blocks	nr	fittings (Note S6)	
	Fishplates	nr		

* *BS.9 is now superseded by BS.11*

COMMENTARY

Supply Only: Rails and other Materials for Plain Track (Refer to Tables 14.01 and 14.03)

Item descriptions for the supply only of rails for plain track give the details as set out against the appropriate descriptive features in Table 14.03 above.

The classification table of Class S lists descriptive features for the supply only of other materials for plain track and provides for them to be given by number. Item descriptions give particulars of incidental fittings and indicate those which are to be included in the items for the featured component. See descriptive features and notes against "Supply only other materials" in Table 14.03.

Where fittings, bearing plates, etc., which qualify to be measured separately are fixed in the manufacture of sleepers, it is convenient to include them in the items for sleepers. Where this is done amending Preamble is given in the Bill of Quantities to the effect that notwithstanding that the CESMM provides for the stated fittings to be measured separately, items for sleepers are deemed to include the stated fittings where it is specifically stated in the item descriptions for sleepers that they are to be included.

Table 14.04   Laying Only: Plain Track

1st Division		2nd Division		3rd Division
Laying only:				
Plain track	Deemed to be for sleepered track laid on loose ballast unless otherwise stated (Note S11)	Plain fishplated track / Plain welded track	m / m	Measure length of completed track (two rail) (Note S12)
	Include lengths of track assemblies in measured lengths (Note S13)	Guard rails	m	Measure length of rail or angle (one rail) (Note S12)
	At junctions measure separate lengths from points of switches (Note S13)	Guard angles / Conductor rails / Crane rails	m / m / m	
	Separate items are not required for bolts, nuts, spikes and rivets used for fixing rails to sleepers or to other foundations (Note S5)	Curved plain track: / radius not exceeding 300 m / radius exceeding 300 m	m / m	Lengths for curved plain track shall not be deducted from the lengths measured for laying plain track (Note S14)

COMMENTARY

Laying Only: Plain Track (refer to Tables 14.01 and 14.04)

The convention adopted for the measurement of laying only plain track are noted in Table 14.04. The lengths given in the Bill of Quantities for items of laying only plain track include that of any curved track and that of any track assemblies. Items in addition are given for curved track and for track assemblies.

Item descriptions for laying only track identify the type of track. It will be taken to be sleepered track laid on loose ballast unless it is otherwise stated.

COMMENTARY

Laying Only: Plain Track (cont.

Item descriptions also identify the type and mass per linear metre of rails, the type of joints and the type of sleepers. Where the intended item coverage of laying only would not otherwise be clear, item coverage Preamble is included in the Bill of Quantities.

Laying only check rails (other than those in switches and crossings), guard rails and conductor rails are measured in linear metres of single rail. Items for these rails make no distinction between straight and curved rails.

Table 14.05   Track Assemblies and Track Accessories

1st Division	2nd Division		3rd Division

Supply only:
 	
Track Assemblies
 	
   Guard rails
   and angles
   in track
   assemblies
   shall not
   be measured
   (Note S12)
 	
Laying only:
 	
Track Assemblies
	Switches and turnouts nr
	
	Diamond crossings nr
	
	Single slip crossings nr
	
	Double slip crossings nr
	
	Other track assemblies nr
		State in item descriptions for supply only of track assemblies the crossing angles and leads and part-iculars of guard rails and guard angles (Note S3)
		
		Separate items are not required for sleepers, chairs, bearing plates, spacer blocks, switch lever boxes, rods and other fittings for track assemblies (Note S7)	

Supply only: Track accessories Laying only: Track accessories	Buffer stops nr Wheel stops nr Lubricators nr		

Track Assemblies and Track Accessories (Refer to Tables 14.01 and 14.05)

The supply only of track assemblies and track accessories are enumerated. Item descriptions state or indicate by specification or drawing reference the type. They provide by statement in the item description or by reference to item coverage Preamble the extent of the associated components intended to be included in the items. See Note S7 of the CESMM.

The laying only of track accessories and track assemblies are enumerated. Incidental components for which separate items do not require to be measured are identified in the descriptions of the track assemblies with which they are associated.

EXAMPLE SE.1

Specimen Bill Items

The Example which follows gives specimen Bill items from a Bill of Quantities for permanent way work. Units of measurement are given against the items. The quantities have been omitted.

EXAMPLE SE.1

PART 3 : PERMANENT WAY

Number	Item Description	Unit	Quantity	Rate	Amount £	p
	EARTHWORKS					
	Excavation of					
E210	Cuttings, topsoil	m3		COMMENTARY		
E230	Cuttings.	m3				
E240	Cuttings, material for disposal.	m3				
	Excavation ancillaries					
E511	Trimming of slopes, natural material other than rock.	m2				
E521	Preparation of surfaces, natural material other than rock; to falls to receive membrane.	m2				
	Filling and compaction					
E624	Embankments, selected excavated material other than topsoil or rock.	m3				
E631	150 mm Thick, excavated topsoil to surfaces inclined at an angle exceeding 10 degrees to the horizontal.	m2				
E635	150 mm Thick, imported natural sand; drainage blanket beneath ballast.	m2				
	Filling ancillaries					
E711	Trimming of slopes, natural material other than rock.	m2				
E721	Preparation of surfaces, sand to falls to receive ballast.	m2				
	Landscaping					
E832	Grass seeding as Specification clause 627; to surfaces inclined at an angle exceeding 10 degrees to the horizontal.	m2				
	(1) To Part 3 Summary			Page total		

COMMENTARY

Earthworks are measured under Class E. See Chapter 7.

Excavated material is deemed to be natural material other than topsoil or rock for re-use, unless otherwise stated in item descriptions.

Items for the preparation of earthwork surfaces are measured where they are to receive PermanentWorks other than earthworks.

Trimming to slopes is measured where the sides of earthworks are trimmed to an angle of inclination exceeding 10 degrees to the horizontal.

Number	Item Description	Unit	Quantity	Rate	Amount	
					£	p
	RAIL TRACK					
	Ballast and track foundations (refer to Preamble clause 27)					
S140.1	Granite bottom ballast.	m3				
S140.2	Granite top ballast (volume of sleepers not deducted).	m3				
S190	"Visqueen" polythene sheeting, 1200 gauge, as Specification clause 1423; membrane beneath sand blanket (measured nett with no allowance for laps).	m2				
	Supply only rails for plain track to BS. 11 (refer to Preamble Clause 28)					
S214	Bullhead rails, Rail Section No. 95RBH, mass 46.95 kg/m.	t				
S225.2	Flat bottom rails, Rail Section No. 113A, mass 56.40 kg/m.	t				
	Supply only, other materials for plain track					
S310	Timber sleepers; softwood impregnated, size 2600 x 250 x 130 mm.	nr				
S320	Prestressed concrete sleepers, BR type F27, with pandrol inserts as Drawing No. S/3 and Specification clause 1521.	nr				
S340	Plain chairs, BR type S1, with chair screws and ferrules and oak keys as Drawing No. S/7 and Specification clause 1528.	nr				
S360	Pandrol baseplates, BR type PAN 8, with coach screws and lockspikes as Drawing No. S/14 and Specification clause 1527.	nr				
S390	Pandrol fastenings as Specification clause 1523.	nr				
	(2) To Part 3 Summary			Page total		

COMMENTARY

Preamble referred to in the items is not given in the Example.

Preamble clause 27 states that the items for ballast are to include for forming the sides to slopes, where required.

Preamble clause 28 notes an amendment to the CESMM. It deletes the reference to BS. 9 and substitutes BS. 11 for bullhead rails.

PART 3 : PERMANENT WAY

Number	Item Description	Unit	Quantity	Rate	Amount £	p
	RAIL TRACK (cont.					
	Supply only other materials for plain track (cont.					
S380.1	Pairs of plain fishplates with fish-bolts as Drawing No. S/5 and Specification clause 1506.	nr				
S380.2	Pairs of junction fishplates with fishbolts as Drawing No. S/6 and Specification clause 1529.	nr				
	Laying only plain track					
S410.1	Plain fishplated track of 95RBH bull-head rails (46.95 kg/m) and timber sleepers.	m				
S410.2	Plain fishplated track of 113A flat bottom rails (56.40 kg/m) and timber sleepers.	m				
S420.1	Plain welded track of 113A flat bottom rails (56.40 kg/m) and concrete sleepers.	m				
S480	Curved plain track, radius exceeding 300 m.	m				
S490	Welded joints between 113A flat bottom rails by Thermit Process, including temporary fishplates. (refer to Preamble clause 29).	nr				
	Supply only track assemblies including all requisite fittings, baseplates, clips, fastenings and timbers, as Specification clauses 1520 - 1530					
S510	Turn outs, vertical design 113A flat bottom rails, BR switch type CV, crossing angle 1 in 9¼, lead 25.025 m (toe/nose) as Drawing No. SC/4.	nr				

COMMENTARY

Welded rail joints are given as separate items in the Example. As this may be considered a departure from the CESMM, the Preamble clause 29 noted in the item description states that in the Bill of Quantities welded rail joints are enumerated and given as separate items.

(3) To Part 3 Summary		Page total			

PART 3 : PERMANENT WAY

Number	Item Description	Unit	Quantity	Rate	Amount	
					£	p
	RAIL TRACK (cont.					
	Supply only track assemblies, including all requisite fittings, baseplates, clips, fastenings and timbers, as Specification clauses 1520 - 1530 (cont.					
S520	Diamond crossings, vertical design 113A flat bottom rails, crossing angle 1 in 6, lead 8.745 m (nose to switch diamond) as Drawing No. SC/6.	nr				
S530	Single slip crossing, inclined design 113A flat bottom rails, angle of slip 1 in 7, lead 10.208 m (knuckle to nose) as Drawing No. SC/7.	nr				
	Laying only track assemblies, comprising 113A flat bottom rails with baseplates, fishplated joints and timber sleepers					
S610	Turn out, switches and crossing as Drawing No. SC/4.	nr				
S620	Diamond crossing as Drawing No. SC/6.	nr				
S630	Single slip crossing as Drawing No.SC/7.	nr				
	Supply only track accessories					
S710	Buffer stop, steel rail and timber sleeper construction as Drawing No. TA4.	nr				
S730	'P & M Universal' rail lubricator, Pammek design with 14 kg grease container as Specification clause 1538.	nr				
	Laying only track accessories					
S810	Buffer stop, as Drawing No. TA/4, mass 2.5 tonnes approximately.	nr				
S830	'P & M' rail Pammek lubricator.	nr				
	(4) To Part 3 Summary			Page total		

The work covered by Class T of the CESMM, includes the lining and securing of tunnels, shafts and other subterranean cavities. Tunnels constructed by cut and cover are excluded from the Class. (See Note T20 of the CESMM). Headings are not specifically mentioned in Class T. The cross reference to them in the "Excludes" to Class E indicates that they are to be measured as provided for tunnels. Pipe laying in headings, tunnels and shafts is specifically excluded from Class T and is included in Classes I, J, K and L.

Table 15.01   Tunnels, Shafts and other Subterranean Cavities

---

Generally – Group items together under appropriate headings to distinguish lengths of tunnels and other parts of tunnelling work having different characteristics (Note T1)

For work under compressed air, state the gauge pressure in stages, the first gauge pressure not exceeding one bar, subsequent stages in increments of 0.4 bars (Note T2)

Class as specified requirements in Class A, the provision, standing by and operation of plant and services associated with the use of compressed air (Note T2)

Tunnels constructed by cut and cover are excluded from this Class. The earthworks, in situ concrete and other components of tunnels so constructed shall be classed appropriately (Note T20)

COMMENTARY

Tunnels, Shafts and other Subterranean Cavities (refer to Table 15.01)

Division of the work to distinguish different lengths of headings, tunnels and shafts, as required by Note T1 of the CESMM, is effected in the Bill of Quantities by grouping items under locational headings. Sub-division beyond that of location to distinguish different characteristics, such as differing ground conditions, working environment and methods of construction, is effected by grouping the related items under identifying sub-headings given under the locational heading.

Compressed Air

Tunnelling work specified to be carried out under compressed air is given separately. It is itemised to distinguish work carried out within different stages of gauge pressures. (See second paragraph of Table 15.01). Item descriptions state, or are listed under headings that state, the gauge pressure applicable to the work to which they relate.

In addition to the items of work measured and given under Class T of the CESMM, items of "specified requirements" are given under Class A for the establishment and removal of compressed air installations and facilities and their operation and maintenance. (See CESMM, Code A278 and Notes A4 and A6). Where different gauge pressures are to be operated for the same length of tunnel, items for operation and maintenance need to be quantified. A suitable unit of measurement is linear metre of tunnel drive, giving separately the quantities for each of the appropriate stated stages of gauge pressures. Any provision for standing by periods is given in items quantified in units of time, i.e. days or weeks.

Table 15.02   Excavation

Generally – Where tunnels are sloping at a gradient of 1 in 25 and steeper, state gradient (Note T3)

Where tunnels or shafts are curved, state radii of curvature (Note T3)

Identify tapered tunnels and shafts (Note T3)

State inclination to the vertical, where shafts are inclined (Note T3)

1st Division

Excavation	State the material to be excavated

2nd Division			3rd Division
Tunnels in rock	m3	Calculate the volume to the payment lines shown on the Drawings or, where no payment lines are shown to the nett dimensions of the volumes to be excavated (Note T5)	State internal diameter where tunnels and shafts are of circular cross-section (Note T7)
Tunnels in other stated material	m3		Where tunnels and shafts are not of circular cross-section substitute maximum dimension of cross-section for diameter and state their cross-section dimensions (Note T7)
Shafts in rock	m3		
Shafts in other stated material	m3		
Other cavities in rock	m3	Class excavation, other than overbreak, outside the normal cross-section profile of tunnels and shafts as excavation of other cavities (Note T5)	
Other cavities in other stated material	m3	Class transitions, breakaways and intersections between tunnels and shafts as other cavities (Note T4)	
Excavated surfaces in rock	m2	Measure area of payment surfaces shown on the Drawings or, where no payment surfaces shown, the net area of the surfaces of the volumes to be excavated (Note T6)	
Excavated surfaces in other stated material	m2	State particulars of filling for voids caused by overbreak (Note T6)	

COMMENTARY

Excavation (refer to Table 15.02)

Excavation of tunnels, shafts and other cavities are each listed as separate classifications in Class T of the CESMM (See Table 15.02). Within each classification, excavation in rock and in any other material, stating the particular material, are each given separately. Material which will qualify to be classified as rock needs to be defined. It is necessary also to define the size at which a boulder, encountered in a volume of soft material, will be classified as rock.

COMMENTARY

Excavation (cont.

      The volume of excavation is given in m3 and is calculated as provided in Note T5 of the CESMM. Some examples of the payment lines for tunnels, established by the rule in Note T5 which obtains where no payment lines are shown on the Contract Drawings, are illustrated in the following diagrams (Figure T1).

      Item descriptions for excavation circular in cross-section state the diameter of the excavation at Third Division level. They state the cross-sectional dimensions of the excavation where it is not circular in cross-section and the maximum dimension of cross-section is used for classification purposes at Third Division level.

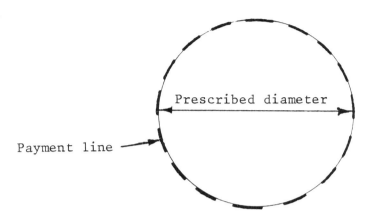

(a) Unlined, where no ground support required

      Disposal of excavated material is not mentioned in Class T of the CESMM. It is taken that the requirements specified for this work is intended to be covered by the items given for the excavation. Any doubt in this respect is avoided by the inclusion of preamble to the effect that the items for excavation of tunnels, shafts and other cavities shall be deemed to include for the disposal of the material excavated as described in the Specification. Separate items with additional descriptions are given for otherwise similar excavation to distinguish different disposal requirements. Where the material excavated is to be brought to the surface for re-use as filling material, this is stated in the item descriptions for the excavation from which the material results. Descriptions for filling items, for which the material is used, indicate it arises from identified tunnelling excavation.

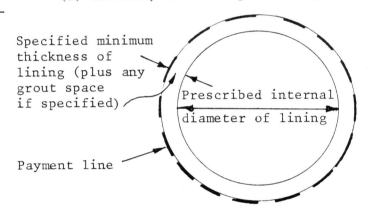

(b) Lined, where no ground support required

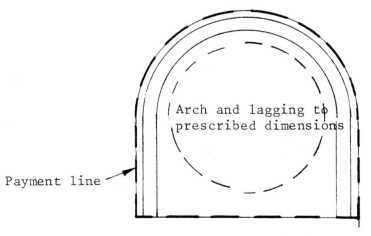

(c) Where ground support required

Fig. T1. Examples of payment lines established by Note T5 of CESMM where none shown on the Drawings.

COMMENTARY

Excavation (cont.

Other Cavities

The application of the classification "other cavities" is given in Notes T4 and T5 of the CESMM. (See notes against the quoted descriptive feature in Table 15.02).

Excavated Surfaces

Items for "excavated surfaces" are measured by area in m2 and are given in addition to those for the volumes of excavation. The areas measured are the net areas of the perimeter surfaces of the excavated voids, assuming the voids to be of the notional cross-section outlined either by the payment lines shown on the Drawings or established from the net dimensions of the volumes to be excavated, as appropriate. Item descriptions state the materials to which the excavated surfaces relate. The filling of voids caused by overbreak is intended to be included in the items. This intention is made clear and particulars of filling to voids is stated in the item descriptions.

Linings (Refer to Table 15.03)

Item descriptions for linings distinguish those to particular configurations of tunnels and shafts as noted against "Generally" in Table 15.03. Classification at Third Division level is by internal diameter where the lining is of circular cross-section. Where not of circular cross-section the maximum dimension of internal cross-section is used for classification purposes and the cross-sectional dimensions are stated in the item descriptions.

In Situ Linings

The thickness measured for in situ linings of designed cross-sectional profile is the net thickness measured from the internal face to the prescribed periphery. In the case of primary linings, the latter is that outlined by payment lines shown on the Drawings, or where no payment lines are shown, that established from the net dimensions of the volumes to be lined.

Item descriptions for in situ concrete linings state or otherwise identify the type and quality of the concrete used and, in the case of sprayed linings the thickness, where they would not be clearly identified by using merely the standard descriptive features. Further details which may require to be given in item descriptions are noted in Table 15.03. The Table gives also the units of measurement. Further Commentary on in situ linings is given in the Measured Example at the end of this Chapter.

Preformed Segmental Linings

Items for preformed segmental linings give the number of complete rings of segments of stated width. Item descriptions list the components which comprise one ring of segments and state the maximum piece weight. See notes against the feature "preformed segmental linings" in Table 15.03. Packing and caulking are measured and given in items separate from those for the segments.

Table 15.03 Linings

Generally   State radii of curvature of curved tunnels and shafts (Note T3)

Identify tapered tunnels and shafts (Note T3)

State gradient of tunnels sloping at a gradient of 1 in 25 or steeper (Note T3)

State inclination to the vertical of inclined shafts (Note T3)

Class transitions, breakaways and intersections between tunnels and shafts as other cavities (Note T4)

1st Division		2nd Division		3rd Division
In situ linings to tunnels In situ linings to shafts In situ linings to other cavities	Measure thickness to payment lines shown on the Drawings or, where no payment lines shown, to the nett dimensions of the volumes lined, see Note T5 (Note T8). State when to head walls, shaft bottoms and other similar components (Note T9)	Cast concrete primary m3 Sprayed concrete primary m2 Cast concrete secondary m3 Sprayed concrete secondary m2 Formwork to cast concrete linings m2		State internal diameter where tunnels, shafts and other cavities are of circular cross-section (Note T7) Where tunnels, shafts and other cavities are not of circular cross-section substitute maximum dimensions of cross-section for diameter and state their cross-section dimensions (Note T7)
Preformed segmental linings to tunnels Preformed segmental linings to shafts Preformed segmental linings to other cavities	Measure number of rings of segments (Note T10) List components which comprise one ring of segments, including number of bolts, grummets and washers (Note T10) State maximum piece weight (Note T10) State when used in pilot tunnels and shafts. Materials to remain property of Employer unless otherwise stated (Note T11)	Precast concrete bolted nr Precast concrete expanded nr	State whether flanged or solid (Note T12)	
		Cast iron bolted nr Cast iron expanded nr Nodular iron nr Fabricated steel nr	State when with machined abutting surfaces (Note T12)	
Preformed segmental linings to tunnels Preformed segmental linings to shafts Preformed segmental linings to other cavities		Lining ancillaries		Parallel circumferential packing nr Tapered circumferential packing nr Stepped junctions nr Caulking m

Table 15.04 Support and Stabilization

1st Division			
Support and stabilization	Measure both temporary and permanent (Note T13)		

2nd Division	3rd Division			
Rock bolts m	Mechanical Mechanical grouted Pre-grouted impacted Chemical end anchor Chemical grouted Chemically filled		State size, type, shank detail, maximum length and number (Note T14)	
Internal supports	Steel arches: supply t erection t			
	Timber supports: supply m3 erection m3		Measure volume as set out in Class O (Note T16)	
	Lagging m2		State materials used for lagging and for packing or grouting behind lagging (Note T15)	
	Sprayed concrete m2 State minimum thickness (Note T16)		Measure area at payment lines shown on Drawings or, where no payment lines shown, to nett dimensions of support to be provided (Note T16)	
	Mesh or link m2		State size and mass of mesh or link fabric (Note T16)	
Pressure grouting	Sets of drilling and grouting plant nr Collaring, securing and providing face packers to holes nr			
	Deep packers of stated size nr		State size	
	Drilling and flushing to stated diameter m Re-drilling and flushing m		State diameter	State lengths of holes in stages of 5 m (Note T17)
	Injection of grout materials of stated composition t		State composition	Exclude mass of water from mass measured (Note T18)
Forward probing m	State length of holes in stages of 5 m (Note T17)			
Stand by driving or sinking operations h	Measure time during which driving and sinking operations are stopped in order to install rock *bolts, to provide internal support, to carry out pressure grouting or forward probing and to carry out any other work classed as support and stabilization (Note T19)			

* *The word "bolts" should be substituted for the the word "belts" in the fourth line of Note T19 in the CESMM by a statement to that effect in the preamble of the Bill of Quantities.*

COMMENTARY

Support and Stabilization (refer to Table 15.04)

The rule in Note T13 of the CESMM requires that both temporary and permanent support and stabilization shall be measured. The Engineer may consider it unnecessary to fully implement the rule where tunnelling in certain stratum is concerned. For soft ground tunnelling it may be considered reasonable to provide that the Contractor is to be responsible for and include in his excavation rates for ground supports and box ups for short periods (such as week ends and public holidays) when driving is not in progress. To provide on these lines is a departure from the CESMM and Bill Preamble needs to indicate that notwithstanding Note T13 of the CESMM temporary support is not measured and the Contractor is to include as stated in the preceding sentence. Measured items should be given for safety bulkheads and where "box up" of the face is required for longer periods than above mentioned.

Where it is anticipated that temporary ground support will be required and it is to be given in the Bill of Quantities as measured items, the Engineer needs to provide drawings and details of his proposals to allow the quantities to be prepared. It is considered that by doing this it does not necessarily commit him to accept responsibility for the design of the temporary support work. The Specification should make clear the extent of the responsibility for the design of the temporary support and stabilization which the Contract places on the Contractor. Where the Contractor has no discretion and must conform to the Engineer's designed requirements, the Engineer is held responsible for the design and the temporary support and stabilization work is admeasured as designed. Where the Specification makes clear that the given drawing and detail are notional and that the Contractor is to be responsible for the design of the temporary support and stabilization, the work admeasured is that actually carried out by the Contractor. Where the latter condition is to obtain, it is sensible for the Contract to require the Contractor to intimate at the time of tender and obtain, prior to the acceptance of the tender, the Engineer's agreement to the use of any system of temporary support which is not envisaged by the proposals in the tender documents. Fresh items and rates can then be agreed for any approved alternative systems.

Items of work classed as "support and stabilization" are listed in Table 15.04. The Table notes the details to be given in the item descriptions of the work and indicates the units of measurement for the various items.

In the absence of any rule in Class T, it is considered reasonable to follow Class M and include the mass of fittings with that of the arch ribs when calculating the mass of steel arches.

Standby Driving and Sinking Operations (refer to Table 15.04)

Items for standby driving operations of tunnels are given separately from those for standby sinking operations of shafts. Item descriptions distinguish different tunnels and different shafts. Items for estimated periods of standby driving or sinking operations are given in the Bill of Quantities quantified in hours. Payment against the items is made for the periods during which the driving or sinking operations are stopped to carry out work classed as support and stabilization. (See Note T19 of the CESMM.) Preamble will usually provide that such payment is conditional upon the incident of carrying out support and stabilization work being such as to make it essential for driving or sinking

COMMENTARY

Standby Driving and Sinking Operations (cont.

operations to stop whilst it is carried out and also that any stoppage of
the operations in such circumstances be notified to the Engineer without
delay. Preamble will also indicate that payment against the standby items
will not be made for periods during which support and stabilization is
carried out as part of the normal cycle of driving or sinking operations,
or for periods when operations are stopped for the boxing up of faces prior
to their being left at the end of work periods.

EXAMPLE TE.2

Measured Example

The measured example which follows includes the sewer tunnels between MH1 - 2
and MH2 - 3 as shown on Drawing No. T/D/1 and one shaft, MH3 as shown on
Drawing No. T/D/2.

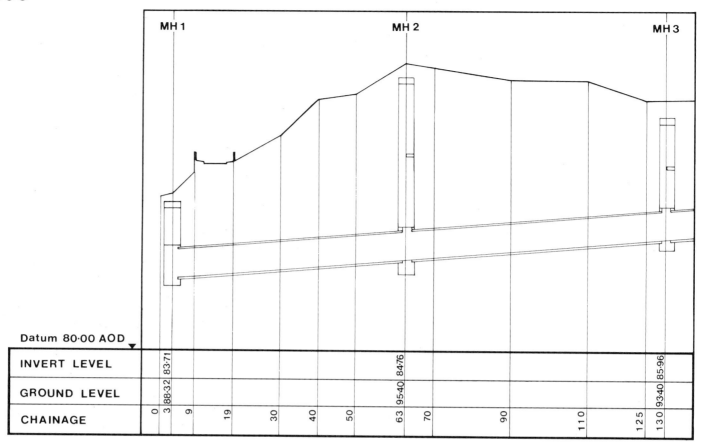

	MH 1						MH 2				MH 3	
Datum 80·00 AOD												
INVERT LEVEL	83·71						84·76				85·96	
GROUND LEVEL	88·32						95·40				93·40	
CHAINAGE	0 3	9	19	30	40	50	63	70	90	110	125	130

SCALES—H 1:1000 & V 1:200 (approx)

LONGITUDINAL SECTION

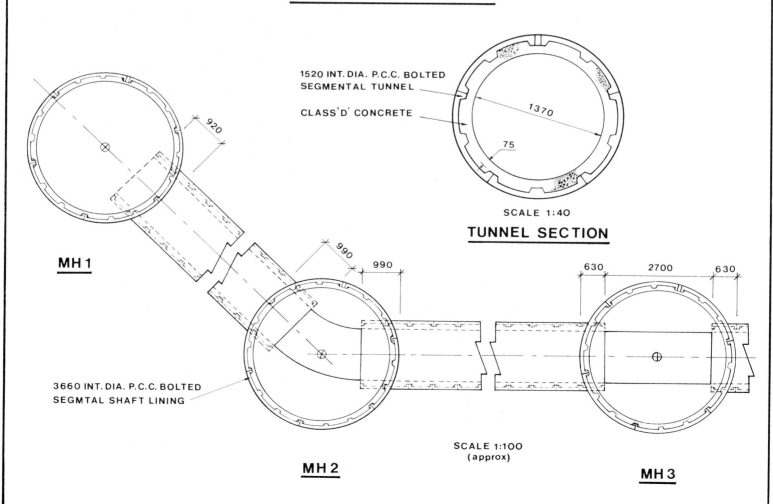

1520 INT. DIA. P.C.C. BOLTED SEGMENTAL TUNNEL

CLASS 'D' CONCRETE

1370

75

SCALE 1:40

TUNNEL SECTION

MH 1

920

990   990

630   2700   630

3660 INT. DIA. P.C.C. BOLTED SEGMTAL SHAFT LINING

SCALE 1:100
(approx)

MH 2

MH 3

SEWER TUNNEL DRG No T/D/1

EXAMPLE TE.1

	Tunnel	Shaft
	1370	3510
	150	150
	1520	3660
	150	150
	110	150
Overall diams	1780	3960

MH 1-2 = 60.00   MH 2-3 = 67.00
2/1/ 3.96 = 3.96   3.96
           56.04   63.04

                    56.04
                    63.04
                   119.08

Tunnels. MH's 1-3

Excavation

$\frac{22}{7}$/ 119.08
0.89
0.89

Tunnels in rock
diam. 1.78m (T111

$\frac{22}{7}$/ 119.08
1.78

Excavd. surfs in
rock, voids filled
with cement grout
as Spec. clause 4.1.
                (T170

1.980 x 1.980 = 3.920
0.890 x 0.890 = 0.792
                3.128

$\sqrt{3.128}$ = 1.769
1.980 - 1.769 = 0.211
$\frac{1.980}{1.769}$ = 1.1193 = Sec 26°40'
26.66 x 2 = 53.33
$\frac{53.33 \times 3.96 \times 22}{360 \times 7}$ = 1.844

(1)

COMMENTARY

The tunnels and shafts in the Example are designed to use standard concrete tunnel segments which are available from several manufacturers. Literature published by the manufacturer of the segments provides data necessary when preparing the quantities.

The two lengths of tunnel in the Example have similar characteristics. The work in the tunnel between MH's 1 and 2 need not, therefore, be distinguished from that between MH's 2 and 3.

The dimensions in the adjoining column are for the excavation and excavated surfaces items between the outer faces of the shafts, leaving the breaking out of the shaft segments and the removal of the natural material at the intersection of the tunnels and shafts to be dealt with in items measured later for "Excavation of other cavities".

The diameter stated in the adjoining item description for excavation is the nominal excavated diameter, i.e. the diameter to the payment lines.

Waste calculations at the foot of the adjoining column are the preliminary calculations for the dimensions of "other cavities" (next column). Using the known dimensions of 1980 mm (shaft radius overall) and 890 mm (half external diameter of tunnel) and the Theorem of Pythagoras the dimension 1769 mm (see diagram below) is calculated. This deducted from 1980 mm gives the 211 mm dimension. The waste subsequently calculates the length of arc which represents the major axis of the elliptical area of segments to be broken out.

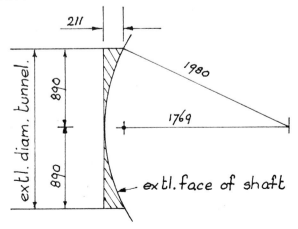

PLAN AT INTERSECTION OF
TUNNEL & SHAFT

SEWER TUNNEL DRG. No. T/D/1

<u>Excavation (cont.</u>

2)1780   2)1844
  890   922

2/2/22/7 / 0.92
0.89
0.11

Other cavities in precast rfcd. conc; breaking away p.c. conc. segments at intersecs of tunnels with shafts (T160

890
922
1812

2/2/22/7 / 1.81
0.11

Excav. surfs. in p.c. rfcd. conc., voids filled with cement grout as Spec. clause 4.1. (T180

8/0 = 0
4/211 = 844
12)844
70

2/2/22/7 / 0.89
0.89
0.07

Other cavities in rock; intersecs. of tunnels and shafts (T150

2/2/22/7 / 1.78
0.07

Excav. surfs in rock, voids filled with cement grout as Spec. clause 4.1 (T170

(2)

The average thickness of the precast concrete shaft segments to be broken out is determined with the aid of the manufacturer's literature, by dividing the volume of a ring, 0.847 m3, by its area 7.59 m2 = 0.112 m average thickness.

The area of the shaft segments to be broken out for the entry of the tunnel into the shaft is that of an ellipse.

The rock to be excavated at each intersection of the tunnels with the shafts and which qualifies to be classified "Other cavities" is indicated hatched in plan on the diagram adjoining the preceding column of dimensions. The dimensions set down for the volume of an intersection is an approximation. They represent the cross-sectional area of the tunnel multiplied by the average of the dimensions shown on the diagram below. This average dimension is used with the circumference of the tunnel to calculate the area of the excavated surfaces of the other cavities in rock. The approximations are considered sufficiently accurate for the particular conditions and the quantities here involved. Where greater accuracy is necessary precise formulae would be used.

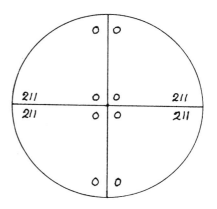

Items for probing ahead and for temporary support for tunnels in rock should be measured in circumstances other than where there is good evidence they are unlikely to be required. The Example assumes the tunnel is in rock of a nature which will not require temporary support to the excavation. Items for these are not, therefore, measured. Preamble would provide that notwithstanding Note 13 of the Class T of the CESMM, temporary support of tunnels is not measured and rates for excavation are to include for temporarily supporting the excavation at all times.

Primary lining

	MH 1 2	MH 2 3
nett lgth of excavn.	56.04	63.04
projecting into MH's {	0.99	0.99
	0.92	0.63
	57.95	64.66

57.95 ÷ 0.61 = 95
64.66 ÷ 0.61 = 106
122.61   201

Preformed segmental linings to tunnels

201

Precast conc. boltd. segmental tunnel linings, flanged, intl. diam. 1.52 m, ring width 0.61 m, comprising 5 segments and key piece, nom. max. piece weight 0.120 t with 20 bolts 40 washers 40 grummets longitudinal jointing strips and circumferential jointing rope (T511

95 - 1 = 94   106 - 1 = 105

{ Lining ancillaries, caulking (T574

94/22/7/	1.52
94/6/	0.61
105/22/7/	1.52
105/6/	0.61

(3)

COMMENTARY

The unit of measurement for segmental tunnel linings is the number of rings of segments. See Note T10 of the CESMM. The number of rings is found by dividing the length of the linings by the ring width of 0.61 m.

The internal diameter stated in the item descriptions for the segmental linings is that to the inner face of the flanges. See Note T7 of the CESMM. The other detail given in the item description for the lining is that which Note T10 of the CESMM provides shall be stated. See Table 15.03. The nominal maximum piece weight may be calculated from the largest piece in a ring of segments. In this case, the manufacturer's literature for the segments, proposed in the Example, gives the total weight for a standard ring of segments. This reduced by an allowance for the key has been divided by five to calculate the nominal maximum piece weight given.

The concrete tunnel segments envisaged in the Example are cast with caulking grooves on the internal circumferential and longitudinal sides. The lining ancillaries item is given for caulking these grooves.

SEWER TUNNEL DRG. NO. T/D/1

Secondary lining

1780	1370
1370	1780
2)410	2)3150
205	1575

Conc. Class "D" as
Spec. clause 4.2

In situ linings to
tunnels

$\frac{22}{7}$ /	122.61	
	0.21	
	1.58	Cast conc. secondary, intl. diam. 1.37m (T23)

201 /	0.26	Ddt { volume { of p.c. { segments
	1.00	
	1.00	

$\frac{22}{7}$ /	122.61	Fmw. fair finish to cast conc. linings, intl. diam 1.37 m. (T25)
	1.37	

COMMENTARY

The Drawing shows the tunnels
to be lined with an in situ
concrete secondary lining,
this is measured in m3, as
required by Class T of the
CESMM.   Item descriptions
state the internal diameter
of the lining.   See Table
15.03.

The thickness inclusive of
the segments and the mean
diameter are calculated in
waste at the head of the
adjoining column.

In the Example, no allowance
is made for creep and the
length of the lining is taken
as that of the segments. See
previous column of dimensions.
The thickness (0.21 m) in the
first set of dimensions for
the concrete lining includes
that of the precast segments.
The second set deducts the
volume of the segments as taken
from the manufacturer's literature.   The volume is given as
0.26 m3 per ring.   This is
entered with two dimensions
each of 1.00 m to indicate it
is a cubic dimension.   The
timesing is the number of rings.

The formwork to the internal
surfaces of the lining is
measured by area.   The item
descriptions state the internal
diameter.   See Table 15.03.

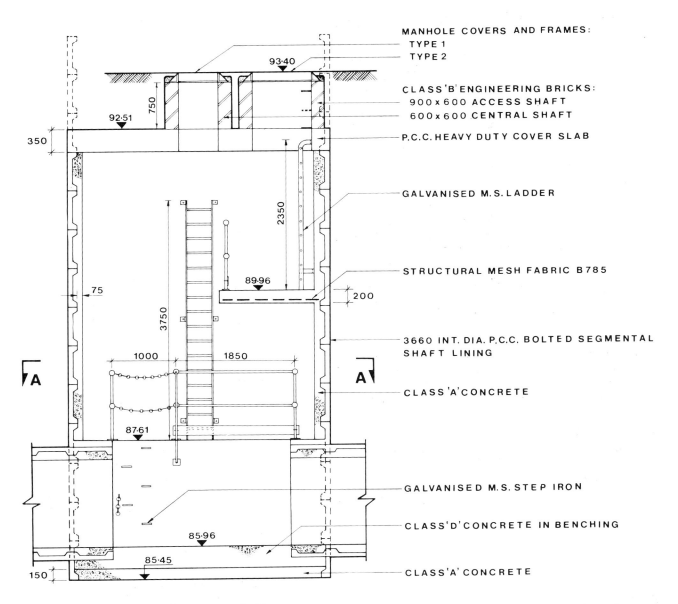

MANHOLE COVERS AND FRAMES:
TYPE 1
TYPE 2

CLASS 'B' ENGINEERING BRICKS:
900 x 600 ACCESS SHAFT
600 x 600 CENTRAL SHAFT

P.C.C. HEAVY DUTY COVER SLAB

GALVANISED M.S. LADDER

STRUCTURAL MESH FABRIC B 785

3660 INT. DIA. P.C.C. BOLTED SEGMENTAL
SHAFT LINING

CLASS 'A' CONCRETE

GALVANISED M.S. STEP IRON

CLASS 'D' CONCRETE IN BENCHING

CLASS 'A' CONCRETE

SECTION B-B

SCALE   0   1   2 m

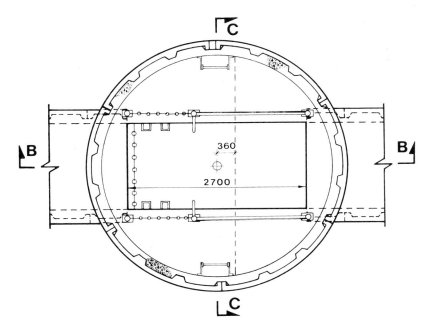

SECTIONAL PLAN A-A

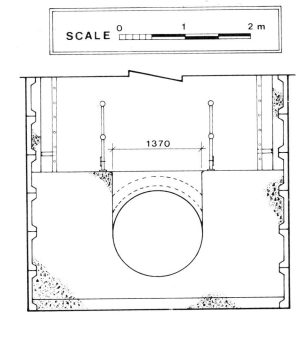

SECTION C-C

MANHOLE No 3 DRG. NO. T/D/2

Extl. diam = 3.96 m
∴ radius = 1.98 m

```
        93.40
        93.10
        0.30
```

Shaft

Excavation

$\dfrac{22}{7}$ / 1.98
   1.98
   0.30

Shafts in topsoil, diam. 3.96 m
(T143.1

```
        93.10
        91.27
        1.83
```

$\dfrac{22}{7}$ / 1.98
   1.98
   1.83

Shafts in made ground, diam 3.96 m.
(T143.2

```
        91.27
        85.45
        5.82
```

$\dfrac{22}{7}$ / 1.98
   1.98
   5.82

Shafts in rock, diam 3.96 m
(T133

$\dfrac{22}{7}$ / 3.96
   0.30

Excavd. surfs. in topsoil (T180.1

```
        92.51      93.10
        0.35       92.16
        92.16      0.94
```

$\dfrac{22}{7}$ / 3.96
   0.94

Excavd. surfs. in made ground (T180.2

(5)

COMMENTARY

The dimensions taken are for one shaft only (Manhole No. 3) as shown on Drawing No. T/D/2.

The calculation of the external diameter of the shaft is given in waste in Column (1) of the dimensions for this Example.

Item descriptions for excavation state the material to be excavated and also the diameter of the shaft. It is assumed the borehole logs would indicate the materials in made ground.

No payment lines are shown on the Drawings and the excavation measured is the nett volumes to be excavated. Depths are taken from reduced levels as follows:-

Ground level	93·40
TOPSOIL	93·10
MADE GROUND	
	91·27
SANDSTONE	
Bottom of shaft	(85·45)

Items for excavated surfaces are measured as provided in Note T6. See Table 15.02. The excavated surfaces in contact with the temporary rings of segments at the top of the shaft are given in separate items from those in contact with the permanent rings because no grouting is required behind the temporary rings. In the bill separate items would distinguish excavated surfaces for shafts from those for tunnels.

MANHOLE NO.3 DRG. NO T/D/2

			1.83
			0.94
			0.89

Excavation (cont.

22/7	3.96	Excavd. surfs. in made ground, voids filled with cement grout as Spec. clause 4.1 (T 180.3
	0.89	

22/7	3.96	Excavd. surfs. in rock, voids filled with cement grout a.b. (T 170.1
	5.82	

22/7	1.98	Excavd. surfs. in rock; shaft bottoms (T 170.2
	1.98	

$$92.16$$
$$85.45$$
$$6.71 \div 0.61 = 11$$

Primary lining
Preformed segmental linings to shafts

11	Precast conc. boltd. segmental shaft linings, flanged, intl. diam. 3.66 m, ring width 0.61 m, comprising 6 segments and key piece, nom. max. piece weight 0.31 t, with 36 bolts, 72 washs. 72 grummets, 7 longitudinal jointing strips and circumferential jointing rope. (T 613

(6)

COMMENTARY

Items for excavated surfaces state particulars of filling for voids caused by overbreak. See Table 15.02.

An item for excavated surfaces is measurable to the bottom of the shaft. To distinguish the work from that to the sides it is given separately. Additional description gives location.

The shaft is designed to use standard concrete tunnel segments. See the first paragraph of the Commentary to Column (1) and the Commentary to Column (3) of the Example.

214 <u>MANHOLE No 3 DRG. No. T/D/2</u>

<u>Primary lining</u> (cont.

10/22/7/ 3.66
11/7/ 0.61

{ <u>Lining ancillaries,</u>
{ caulking (T574.2

<u>General items</u>
<u>Specified reqmts.</u>
<u>Temp. Works</u>
<u>Safety measures to</u>
<u>top of Shafts</u>

<u>sum</u>

Establishment and removal of three temporary rings of precast conc. segments, intl diam. 3.66 m, ring width 0.61m to top of MH.3. as shown on Drawing No T/D/2 and in accordance with Spec. clause 3. (excav. measd. sep.) (A279.1

The Example assumes an express requirement for three temporary rings of shaft segments to be erected for safety reasons at the top of the shaft. These are shown on Drawing No. T/D/2. In accordance with Note A3 of the CESMM, these Temporary Works are given as Specified Requirements. Items which separate establishment and removal from operation or maintainance are given as required by Note A6.

The item description for establishment makes clear that the excavation for the temporary rings is measured separately.

<u>sum</u>

Maintenance of temporary rings of segments to top of MH.3. until cover slab fixed and in position (A279.2

(7)

Secondary linings

U/s cov. slab.	92.16
Top of benchg.	87.61
	4.55

3960	3510
3510	3960
2) 450	2) 7470
225	3735

In situ linings to shafts

22/7 /	3.74	Cast conc. second-
	0.23	ary, diam. 3.51 m;
	4.55	conc. Class "A" as
		Spec. clause 4.2 (T333

0.61) 4.55
7.46

7½/	0.75	Ddt ditto {volume
	1.00	{of p.c.
	1.00	{segments

22/7 /	3.51	Fmw. fair finish to
	4.55	cast conc. linings,
		intl. diam. 3.51 m

3.66   (T353
3.81
2) 7.47
3.74 ÷ 2 = 1.87

22/7 /	1.87	Cast conc. primary
	1.87	to form shaft bottom.
	0.15	conc. Class "A" a.b.
		(T310

Benching
87.61
2.16	85.45
0.15	2.16
2.01	

2) 3.96
1.98

COMMENTARY

To simplify measurement, the thickness of the in situ concrete lining, entered in the first set of dimensions in the adjoining column, includes that of the precast concrete segments. The next set of dimensions deducts the volume of the segments (0.75 m3 per ring) as taken from the manufacturer's literature. The timesing factor is the number of rings. The height of the in situ lining is measured from the top of the benching to the underside of the precast concrete cover slab. This is equal in height to 7.46 rings of segments.

The in situ lining to the bottom of the shaft is a primary lining and the description in the Example classifies it accordingly. As required by Note T9 of the CESMM the item description states it is to shaft bottoms.

MANHOLE No 3 DRG. No. T/D/2

<u>Benching</u> (cont.

$\frac{22}{7}$/	1.98	Provn. of conc.
	1.98	Class "D" as Spec.
	<u>2.01</u>	clause 4.2 (F243.2

&

Placg. of conc., mass benchings to m.h. bottom 3.81 m diam x 2.01 m (extreme)   (F680

	2.70	{Ddt. Both last items
	1.37	
	0.97	(main channel
$\frac{1}{2}$/$\frac{22}{7}$/	2.70	630 – 211 = 419
	0.69	630
	0.69	(ditto. „ <u>419</u>
$\frac{2}{22}$/$\frac{22}{7}$/	0.89	3)1468
	0.89	<u>489</u>
	<u>0.49</u>	(projecting ends of {tunnels
$3\frac{1}{2}$/	0.75	
	1.00	(volume of p.c.
	1.00	(segments.

$\frac{2}{22}$/$\frac{22}{7}$/	0.92	{Add. Both last items
	0.89	(segments over
	0.11	(deducted previously
$\frac{22}{7}$/	3.74	1.65
	0.15	fall. <u>0.02</u>
	0.15	1.67
	<u>1</u>	Fmw. fair finish to intl. surfs. of in situ conc. "U" shaped channel, length 2.70m, max. dimensions of cross-section 1.37 x 1.67 m, inc. fmw. to ends (G290

(9)

COMMENTARY

The excavation and lining of the shaft has been classified in accordance with Class T. Internal work in the shaft is given in accordance with other Classes.

The in situ concrete benching is measured by volume as provided in Class F. Measurement of the benching is simplified in the Example, by first including the precast segments in the volume measured and afterwards deducting their volume on the basis of that given in the manufacturer's literature.

The "Add" item which follows the "Ddt" in the adjoining column of dimensions adds the two items of concrete over deducted by (i) the tunnel projecting into the shaft, and (ii) the flange of the shaft segments extending into the bottom lining.

Because the section of the channel changes due to the fall in the invert it is inappropriate to measure it as of constant cross-section. It is given in the Example as a special numbered item in preference to measuring it by area.

MANHOLE No 3 DRG No T/D/2

Benching (cont.

22/7/	1.76	
	1.76	

Conc. accessories
fin of top surfs.,
steel trow. fin. (G812

	2.70
	1.37

Ddt ditto (channel

$$2\,)\underline{1.370} \qquad \begin{array}{r} 1.650 \\ 0.685 \\ \hline 0.965 \end{array}$$

2/	2.70
	0.97
½/ 22/	2.70
2/ 7/	1.37
2/½/22/	0.69
7/	0.69
2/	1.37
	0.97

Conc. accessories
fin. of formed surfs
of "U" shaped channel
as Spec. clause (G823
                 (ends
                 (ends

2/22/	0.69
7/	0.69

Ddt Last item (tunnel
            (openings

Intermediate slab

½/22/	1.76
7/	1.76
	0.20

Provn. of conc.
Class "A" as Spec.
clause 4.2   (F243.1

+

Placg. of conc., rfcd.
susp. slabs, thickn
150-300 mm (act 200);
inter slab. (F732

$$2\sqrt{1755^2-360^2} = \begin{array}{r} 3510 \\ 3435 \\ \hline 2\,)\,6945 \\ \hline 3473 \end{array}$$

	3.47
	0.36
	0.20

Ddt. Both last items

(10)

COMMENTARY

Items for finishing surfaces
are given where a separate
finishing treatment to the
surface of the concrete is
required. The Example
assumes a specified treatment to the formed surface
of the channel after the
removal of the formwork.

The intermediate concrete
landing is measured as
provided in Class F ot the
CESMM. The plan shape of
the landing is segmental.
In the Example the dimensions set down for it are
those for a semi-circle
followed by the deduct of
the want to reduce the
semi-circle to the required
segment. See diagram below.

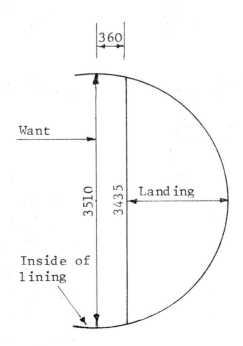

MANHOLE No. 3 DRG No. T/D/2

<u>Inter. slab (cont.</u>

$$\frac{\pi\ 3.51^2}{8} = 4.840$$

<u>Less</u> $3.47 \times 0.36 = 1.249$

$3.435) \overline{3.591}$

av. width landing = 1.045

$\frac{1}{2}$/$\frac{22}{7}$/	1.76		Fmw. fair finish plane horiz. width 1.05 m (average); soffit of intermed. landing slab, curved cut ag. shaft lining (G214	
	1.76			

&

Rfcmt h.y. steel fabric to BS. 4483 ref. B785 nom. mass 8.14 Kg/m2; in inter landg slab curved cut ag. shaft (measd. nett) (G558

&

Conc. accessories, fin af top surfs., steel trow. fin (G812

3.47		Ddt. Last three items	
0.36			

<u>Exposed edge</u>

3.44	Fmw. fair finish, plane vert, width 0.1 - 0.2 (act. 200) (G242	

COMMENTARY

The formwork to the soffit of the intermediate landing slab is measured by area and is classified plane horizontal stating an average width. Because of its shape it is thought reasonable to provide additional description which includes location. This applies also to the item description for the reinforcement.

The Example assumes that the soffit and edge surfaces of the concrete landing are to be left fair from the formwork and that no separate finishing treatment to the surface of the concrete is required.

The measurement convention of first measuring a semi-circle and deducting a want to reduce it to a segment is adopted for the measurement of the formwork and the reinforcement.

The average width for the classification of the form-work to the soffit of the landing is found by dividing the area by the length on the front edge. See waste at the top of the adjoining column.

The formwork to the edge of the landing being of a width not exceeding 200 mm is measured in linear metres.

MANHOLE NO. 3 DRG. NO. T/D/2

Cover

Precast conc. (cov. slabs.

1 | Heavy duty cover slab to shaft, diam 3.96 m (overall), thickn. 350 mm in four pieces as Drawing No. 5/3 (H530

```
2/2/600=2400   2/900=1800
4/215= 860    2/600·1200
      3260          3000
  93.40      4/215    860
  92.51              3860
  0.89        0.89
Less cover    0.14
              0.75
```

BKw. (access shafts.

3.86
0.75
3.26
0.75

Engin. bkw. as Spec clause 6.12., One bk. constructn., vert walls; access shafts (U311

3.00
0.75
2.40
0.75

Engin. bkw. Surf features fair facg. (U378

Misc. Metalwk

1 | Cast iron coated M.H. cover and fra, type 1 as Spec. clause 2.39 and Drg No 4/21 (N499.1

1 | Cast iron coatd MH cov + fra, type 2 as Spec. clause 2.39 and Drg No 4/22 (N499.2

(12)

Precast concrete cover slabs, either standard or manufactured to drawn and specified requirements, are available from the manufacturers of precast segments. The cover slab here measured is assumed to be Contractor designed to suit stated dimensions and loading requirements. It is classified under Class H. The mass per unit will depend on the Contractor's design and is not stated in the item description. The zero in the Code number signifies that the Third Division descriptive features in Class H are not appropriate to the item.

The engineering brickwork enclosing the access shafts is classified vertical walls under Class U. Inner faces of the work are taken in the Example to be finished to a fair face. Step irons in the brickwork are measured later with the items of miscellaneous metalwork.

Miscellaneous metalwork is measured under Class N of the CESMM. Code N4 is used in the Example as the First Division code for miscellaneous metalwork components in coated cast iron. See Note N3 of the CESMM.

220

MANHOLE NO 3. DRG. NO.T/D/2

Intl. components

Misc. metalwork
Galv. m. steel step
irons as Spec. clause
2.40, type 2 (cast-
ing in to conc.
measured sep)(N199.1

2/ 5

&

Conc. ancills. Inserts
Fixg. & casting in to
conc. m. h. step irons
and making good fair
formed fin. (G832

Misc. metalwork
Galv. m. steel
handrails as Spec.
clause 2.56 and Drg.
No. 23 inc. mortices
in conc. for fixings
Handrails of solid
steel sections, as
detail "J", length 3.35m
x 1.10m high; inter
landing.   (N199.3

1

2

Handrails of solid
steel sections & with
safety chain, as detail
"K", length 2.85m over-
all x 1.00m high; side
of channel. (N199.4

1

Misc. metalwork
Galv m. steel safety
chain as Spec. clause
2.55, length 1.45m inc.
morts in conc. for
fixings (13) (N199.5

MANHOLE NO 3 DRG. NO. T/D/2

			Intl. components

Misc. metalwork
Galv. m. steel ladders
as Spec. clause 2.57
and Drg. No. 26
inc. mortices in conc.
for fixings

2/	2.35		{Ladders (measured
	3.75		{vertically) (In No.3)
			(N130

Misc. metalwork
Galv. m. steel step
irons as Spec. clause
2.40, type 1 and
buildg. in to engin.
bkw. (N199.2

| 3/ | 1 | | |

Backfill at top

	93.40	0.89
	92.51	0.15
	0.89	0.74

Earthworks
Fillg. and compact. to
structures, imported
fillg. matl., type 'C'
as Spec. clause (E615

22/	1.98		
7	1.98		
	0.74		

	600	900
2/215	430	430
	1030	1330

	1.03		{Ddt. Last item
	1.03		
	0.74		access shaft
	1.33		
	1.03		(" "
	0.74		

(14)

Ladders are here measured in linear metres as required by Class N of the CESMM. It is considered of help to give the number of ladders and the convention used in measuring them as additional description in the item.

Whilst the building in to brickwork of such items as pipes and ducts are required by Class U of the CESMM to be given as separate items, it is thought reasonable to include the building in with the item for the step irons.

The Example assumes that the Specification requires that unless otherwise specified, excavated material is for disposal. The Example assumes that the excavated material is unsuitable for filling. Filling material is taken as imported. Had it been required that excavated material be used for filling it would have been classified accordingly and a volume of excavation of shaft equal to that required for filling would have been re-classified as for re-use.

MANHOLE No 3. DRG No T/D/2

<u>Reinstate at top</u>.

<u>Earthworks</u>

22/7	1.98	
	1.98	

Fillg & compact, thickn 150 mm, imported top soil; making out grassland over shafts

(E 632

&

Landscapg. grass seedg., surf n.e. 10 degrees, Spec. clause 3.9; making out grass land over shafts

(E 831

1.03	
1.03	Ddt. Both last items
1.33	(access shaft
1.03	(" "

COMMENTARY

The Example assumes that there is no express requirement for the preservation of topsoil. The adjoining item for reinstatement provides for topsoil to be imported. Had there been an express requirement to preserve the topsoil, the item description for "Excavation, shafts in topsoil" (Column 6 of the Example) would have stated an Excavate Surface of "underside of topsoil". Also the remaining items for the excavation of shafts would have stated a Commencing Surface of "underside of topsoil". Additionally, items for excavation of shafts in topsoil would have distinguished that for disposal from that for re-use, if necessary.

(15)

16 Brickwork, Blockwork and Masonry—CESMM Class: U

Brickwork in manholes and brickwork incidental to pipework are excluded from Class U and are included in Class K of the CESMM. It is appropriate to measure and classify all other brickwork and all blockwork and masonry in accordance with Class U. The classification table for this Class provides a separate set of descriptive features for each of the materials. The rules of measurement and the descriptive features are the same for each material, with minor exceptions.

Table 16.01 Brickwork, Blockwork and Masonry

Generally - Include the volumes and areas of joints in the volumes and areas measured (Note U5)

No deduction or addition for rebates or projections less than 0.05 m2 cross-sectional area. No deduction for holes and openings less than 0.25 m2 cross-sectional area (Note U5)

Separate items are not required for joints, pointing, fixings or ties (Note U3)

COMMENTARY

Brickwork, Blockwork and Masonry (refer to Tables 16.01 - 16.04)

Class U provides for brickwork, blockwork or masonry to be given by volume or area, dependent on the thickness of the work, and lists thickness classifications. Within each thickness classification the work is further classified at Third Division level. Further items are given for any surface features and also for any work classified as ancillaries. (See Class U classification table of CESMM and the Tables given in this Chapter).

Sinkings and projections each less than 0.05 m2 cross-sectional area and openings and voids each less than 0.25 m2 cross-sectional area are neglected when measuring the volumes or areas of the brickwork, blockwork or masonry. Otherwise, the volumes or areas given are those of the work; inclusive of the volumes and areas of the mortar joints. (See subsequent Commentary on "Surface Features").

Brickwork and Blockwork (refer to Tables 16.01, 16.02 and 16.03)

Descriptive features for brickwork, blockwork and applicable Notes from the CESMM, are summarised in Tables 16.02 and 16.03, respectively. Within given thickness ranges, work exceeding one metre in thickness is given by volume in m3; that not exceeding one metre in thickness is given in m2 stating the actual thickness. The descriptive feature "One brick construction" applies to brickwork of a thickness equal to the length of the bricks specified. "Half brick construction" applies to brickwork of a thickness equal to the width of the bricks specified. The descriptive feature "One block con-struction" applies to blockwork of a thickness equal to the thickness of the blocks specified.

Table 16.02 Brickwork

1st Division	2nd Division			3rd Division	
Common brickwork Facing brickwork Engineering brickwork Materials, dimensions and types of bricks or BS. reference to be stated (Note U1)	One brick construction	m2	Measure on centre line (Note U5) State thickness (Note U1)	Vertical walls Battered walls Walls battered one side (Note U6) Curved walls Piers, columns and stacks Facing to other materials Mass work Arches	Different bonding patterns for same material to be stated (Note U4) State if in cavity construction – measure each skin separately (Note U4) State if in composite construction – measure separately each layer or skin of different material (Note U4)
	Half brick construction	m2			
	Mass brickwork thickness not exceeding 1 m	m2			
	Mass brickwork thickness 1 – 2 m	m3			
	Mass brickwork thickness 2 – 3 m	m3			
	Mass brickwork thickness exceeding 3 m	m3			
	Surface features	Include sufficient detail to identify special or cut bricks or blocks (Note U7) State spacing of intermittent surface features (Note U7) Reveals are not classed as surface features (Note U7)		Copings and sills m Rebates and chases m Cornices m Band courses m Corbels m Pilasters m Plinths m Fair facing m2	
	Brickwork ancillaries	Joint reinforcement m Damp proof courses m		State materials and dimensions (Note U8)	
		Bonds to existing work m2			
		Concrete infills m2		State thickness State mix specification of concrete (Note U8)	
		Built in pipes and ducts nr		State cross-sectional area range as 3rd Division features	
		Centering to arches m2		Measure temporarily supported intrados (Note U9)	

Table 16.03 Blockwork

1st Division	2nd Division		3rd Division
Lightweight blockwork Dense concrete blockwork Artificial stone blockwork Materials, dimensions and types of blocks or BS. reference to be stated (Note U1)	One block construction m2	Measure on centre line (Note U5) State thickness (Note U1)	The 3rd Division descriptive features and the Notes refer-red to in Table 16.02 for Brickwork are the same for Block-work
	Mass blockwork thickness not exceeding 1 m m2		
	Mass blockwork thickness 1-2 m m3		
	Mass blockwork thickness 2-3 m m3		
	Mass blockwork thickness exceeding 3 m m3		
	Surface features	The Notes referred to against this descriptive feature in Table 16.02 for Brickwork are the same for Blockwork	The 3rd Division descriptive features, the units of measure-ment and the Notes referred to in Table 16.02 for Brickwork are the same for Blockwork
	Blockwork ancillaries		

COMMENTARY

Brickwork and Blockwork (cont.

Materials and types and sizes of bricks and blocks, or an equivalent B.S. reference, are stated in the item descriptions for the brickwork or blockwork. Item descriptions for brickwork or blockwork specified to be built in special bricks or blocks describe the bricks or blocks as stock pattern or purpose made, as appropriate.

Masonry (refer to Tables 16.01 and 16.04)

Mass masonry of a thickness not exceeding one metre and single skin masonry are both measured by area in m2. Mass masonry exceeding one metre in thickness is measured by volume in m3. The actual thickness of the masonry is stated in item descriptions irrespective of whether it is measured by area or volume.

Item descriptions for masonry state the materials. Separate items are not required for "fair facing" in respect of masonry. The surface finish of the stone and jointing and pointing requirements are identified in the item descrip-tions for the masonry. Separate items are not required for fixings and ties. Item descriptions for masonry make clear the extent of the metalwork or other fixings and ties intended to be included in the items for the masonry and state or otherwise identify the types of fixings and ties and the materials to be used. Where their inclusion would complicate or unduly extend the extent of itemisation

COMMENTARY

Masonry (cont.

of the masonry it is preferable to give fixings and ties separately.   This
entails a note in Preamble amending the CESMM and a statement in the item
descriptions of the masonry that fixings and ties are measured separately.

Table 16.04   Masonry

1st Division	2nd Division				3rd Division
Ashlar masonry Rubble masonry (State materials (Note U2))	Walls or facework in single skin construction m2 Mass masonry thickness not exceeding 1 m m2	Measure at centre lines (Note U5)	State thickness (Note U2)		The 3rd Division descriptive features and the Notes referred to in Table 16.02 for Brickwork are the same for Masonry

	Mass masonry:
	
	  thickness 1-2 m m3
	
	  thickness 2-3 m m3
	
	  thickness exceeding 3m m3				

	Surface features	The Notes referred to against this descriptive feature in Table 16.02 for Brickwork are the same for Masonry			The descriptive feature "Fair facing" is not applicable to Masonry. Otherwise the 3rd Division descriptive features, the units of measurement and the Notes referred to in Table 16.02 for Brickwork are the same for Masonry.
	Masonry ancillaries				

COMMENTARY

Cavity and Composite Construction

Each skin of cavity construction and each skin or layer of composite construction
is measured separately. For cavity construction it is unnecessary to give separate
items for wall ties, but wall tie requirements should be identifiable in relation
to the cavity construction. Separate items are given for similar skins of cavity
construction to distinguish different tying arrangements. In composite construction where a skin or layer is of varying thickness, it may be convenient to
measure each thickness separately. Where it is not, item descriptions state the
proportions of the various thicknesses or an average thickness; they may state
both where it is considered it would be helpful to do so. See Specimen Item
Descriptions given subsequently.

COMMENTARY

Cavity and Composite Construction (cont.

Number	Item description	Unit
	BRICKWORK, BLOCKWORK AND MASONRY	
	Engineering brickwork in Class B bricks in group 7 mortar	
U391.1	Three quarter brick (average) construction, comprising 50% half brick and 50% one brick construction, vertical walls, composite construction, English bond.	m2
	Engineering brickwork in Class A bricks in group 7 mortar	
U391.2	Three quarter brick (average) construction, comprising 50% half brick and 50% one brick construction, vertical walls, composite construction, English bond	m2
	Surface features	
U378	Fair facing as Specification clause 16.17	m2

SECTION

Legend

Class A bricks faced and pointed

Class B bricks no pointing

Fig. U1. Specimen item descriptions for brickwork composite construction.

Surface Features

In addition to giving items for the volumes or areas of the brickwork, blockwork or masonry, items are given for any surface features in the work. Standard descriptive features are listed in the Class U classification table of the CESMM. (See Table 16.02).

The volumes or areas of the projecting or sunk parts of surface features which are each less than 0.05 m2 cross-sectional area are neglected when measuring the volumes or areas of the items for the brickwork, blockwork or masonry. Otherwise the actual volumes or areas of the work in any surface features are measured as part of the volumes or areas given for the items of brickwork, blockwork or masonry. The additional items given for surface features are then for the work forming the features which (although not described as such) is extra over that given in the items for the volumes or areas of the brickwork, blockwork or masonry in which the surface features occur. This is the usual convention but circumstances may arise which make it preferable for surface features to be given as full value items. Preamble should make clear whether or not the volumes or areas of the work in surface features are included in the volumes or areas given for the work in which they occur.

COMMENTARY

Surface Features (cont.

Number	Item description	Unit
	BRICKWORK, BLOCKWORK AND MASONRY	
	Rubble masonry in Yorkshire stone in cement mortar 1:4, as Specification clause 18.23	
U831	Mass masonry, thickness 400 mm, vertical walls; fair exposed faces both sides.	m2
	Surface features	
U871	Copings; 500 x 100 mm of ashlar masonry to Specification clause 19.14, as Drawing No. U/1.	m
U874	Flush horizontal bands; 100 x 200 mm of ashlar masonry to Specification clause 19.14.	m

DRAWING NO. U/1

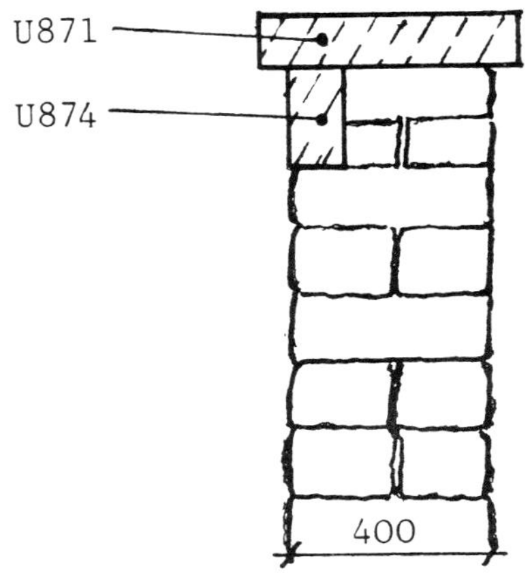

U871

U874

400

SECTION

Fig. U2.   Specimen item descriptions for rubble masonry wall with ashlar masonry surface features.

Consistent with the usual convention that surface features are given as additional cost items, extra over the construction in which they occur, the area given for specimen item Code U831 above would be calculated using the height measured to the top of the coping overall the flush band. The specimen items for the ashlar masonry surface features are given under the classification "Rubble masonry" to denote they occur in the volumes or areas of the rubble masonry and are, in effect, extra over the rubble masonry. Preamble would be included in the Bill of Quantities to make clear the measurement conventions adopted.

A situation where it is thought preferable to give the surface feature of the coping as a full value item is illustrated in Figure U3. Rather than include the area of the coping in that of the single skin construction in which it occurs, it is treated as a special case and given as a full value item. The phrase "full value" is added to the item description for the coping. Preamble would be given stating that the phrase "(full value)" where used in the item descriptions for surface features indicates that the volumes or areas of the work in such features is not included in the volumes or areas given for the brickwork, blockwork or masonry in which such surface features occur.

COMMENTARY

Surface Features (cont.

Number	Item description	Unit
	BRICKWORK, BLOCKWORK AND MASONRY	
	Rubble masonry in Yorkshire stone, as Specification clause 18.22 in group 8 mortar	
U815	Single skin construction, thickness 200 mm nominal with bond stones thickness 325 mm to 15 per cent of face area, facework to in situ concrete walls.	m2
	Surface features	
U871	Copings; 700 x 200 mm in ashlar masonry, as Drawing No. U/3 (full value).	m
U877	Plinths; 50 mm projection x 450 mm high, finished with 250 x 175 mm ashlar masonry chamfered plinth course, as Drawing No. U/3.	m
	Masonry ancillaries	
	"Hyload" pitch polymer damp proof courses (measured nett with no allowance for laps)	
U882	250 mm wide.	m

DRAWING NO. U/3

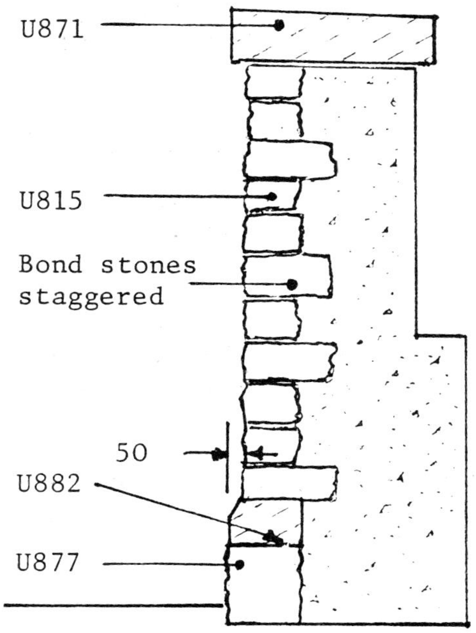

SECTION

Fig. U3.   Specimen item descriptions for rubble faced concrete wall

The specimen item descriptions in Figure U3 do not include those for the concrete wall. The concrete would be measured in accordance with Class F. The item for placing the concrete would indicate it was placed behind single skin masonry construction. Formwork to the back of the wall would be measured as provided in Class G.

EXAMPLE UE.1

Measured Example

A measured example for a brickwork wall with surface features follows.

A-A

113
328
440
2475
750

B-B

113 450
225
2900 RAD.
2325

PLAN

B
B
665
328
2944
328
2944
2718
328
2944
328
665
A
A

NOTE
All brickwork to be in class 'B' engineering
bricks, fair face on exposed surfaces.

BRICKWORK WALL   DRG. No U/D/1
EXAMPLE UE.1

Engin. bkw. in Class B bks. as Spec. clause 3.4 in ct. mor. 1:3 unless o/w described

	Wall.
750	
2475	3/2944 = 8832
113	4/ 328 = 1312
3338	10144

10.14	
3.34	

Mass bkw, thickn. 328 mm, vert walls; English bond. (U331.1

Plinth

2944
2/328   656
3600

2/ 3.60	
0.75	

Ddt. last item (U331.1

&

Add Mass bkw, thickn. 440 mm, vert walls; English bond. (U331.2

Piers

450		3338
113		563
563		2775
	4/665 = 2660	
	4/103 = 412	
	2248	

2/ 2.25	
2.78	

H. bk. constrn. facg. to conc. cols., wi metal anchors as Spec. clause 3.19; stretcher bond. (U325

2/ 0.67	
0.56	

Mass bkw, thickn 665 mm, piers and columns; English bond. (U334

(1)

COMMENTARY

The first item in the adjoining column is that of the mass brickwork in the wall. Its length is measured between the end piers over the opening. The dimensions for the height includes that of the coping. The whole of the wall is first measured as 328 mm thick. Plinth, opening and arch are dealt with as adjustments later.

The brickwork does not exceed one metre in thickness and is measured by area stating the thickness of the work.

The projection of the plinth 112 x 750 mm is greater than 0.05 m2 cross-sectional area. Its volume or area is, therefore, included in the volume or area of the work in which it occurs. The items which deduct 328 mm thick work and adds that which is 440 mm thick adjusts the brickwork for the increased thickness of the plinth at the base of the wall each side of the opening.

The drawing indicates that the brickwork to the lower part of the end piers is facing to concrete columns. Above the columns the piers are in solid brickwork. It is assumed that the concrete is not infill but concrete columns cast using formwork prior to the commencement of the brickwork and that the brick facing is tied with brick anchors to dovetailed slots cast in the concrete. The concrete, the formwork, dovetailed slots and other work in the concrete columns are not measured in the Example. Dovetailed slots cast in to concrete are classed as inserts and are measured under Class G of the CESMM.

Additional description makes clear the tying requirements for the half brick thick facing. Separate items are not required for ties (anchors).

BRICKWORK WALL   DRG NO. U/D/1

		Surfs. features in Class B bks. a.b.
3/	2.94	Copings; bk. on edge, width 328 mm (U371.1
4/	0.33	Copings; bk. on edge, width 440 mm. (U371.2
2/	0.67	Copings; bk. on edge, width 665 mm (U371.3
2/	2.94	Plinths; c.s.a. 0.05 m2 or greater, wi 2 cos. s. patt. splyd. plinth bks. (U377
4/	2.48	Pilasters; 113 mm proj. x 328 mm wide (U376

3338	3338	2660
150	750	328
3188	2588	2332

2/	10.14	{ Fair facing as Spec. clause 3.14 (U378
2/2/	3.19	
	0.11	(pilaster returns
2/2/	3.19	
	0.11	(" "
	2.59	(plinth splay
2/	2.94	
	0.04	(top of coping
3/	2.94	
	0.33	(" " "
4/	0.33	
	0.44	(faces of piers
2/	2.33	
	3.19	(top pier coping
2/	0.67	
	0.67	

(2)

COMMENTARY

The area of the copings having been included in that of the brickwork measured by area, the surface feature items for copings represent additional cost items extra over the brickwork in which they occur. Separate items are given which distinguish different widths of coping because of their different cost characteristics.

The item description for the plinth includes "sufficient detail to identify special or cut bricks". See Note U7 of the CESMM. Projections of 0.05 m2 cross-sectional area or greater are measured as part of the volume or areas of the brickwork. Additional description given in the surface feature item for the plinth indicates a cross-sectional area of 0.05 m2 or greater. From this the estimator will recognise that the surface feature item for the plinth is to cover only the extra forming the plinth and the extra for the special plinth bricks.

The cross-sectional area of the pilasters are less than 0.05 m2. The brickwork in them is not, therefore, included in the brickwork item given by area. Cross-sectional dimensions are given in the surface feature item to provide the information for the estimator to include for the extra brickwork as well as the extra in forming the pilasters. It is helpful to add the phrase (full value) to the description of such items. See Commentary on Surface Features given previously.

The surface features item "Fair facing" is measured to the exposed surfaces of the brickwork including the surfaces of other surface features. The work is measured overall leaving the opening to be adjusted later in the Example.

BRICKWORK WALL   DRG No.U/D/1

Bkw. ancills.

Pitch polymer damp proof courses

2/	2.25	Width 103mm (U382.1
	2.94	Width 328 mm (U382.2
2/	3.60	Width 440mm (U382.3

Arch + Opg.

$\frac{1}{2}/2718 = \underline{1359}$

$\frac{1359}{2900} = 0.4686 = \underline{Sin\ 28°}$

$(Cos\ 28°)\ 0.88295 \times 2900 = \begin{array}{r} 2900 \\ \underline{2561} \end{array}$

rise of arch $\underline{339}$

$\frac{1}{2}/328 = \begin{array}{r} 2900 \\ \underline{163} \\ \underline{3063} \end{array}$

$28° \times 2 = \underline{56°}$

arch soffit = $56 \times 2.90 \times 0.017453 = \underline{2.83}$

arch = $56 \times 3.06 \times 0.017453 = \underline{2.99}$

Engin. bkw. a.b.

	2.99	mass bkw., thickn 328
	0.33	mm, arches; segmental
		in h.bk. rings (U337

Bkw. ancills.

	2.83	Centering to arches;
	0.33	segmental (U388

(3)

COMMENTARY

In keeping with general philosophy of the CESMM the damp proof course is measured nett with no allowance for laps. This would be made clear in Preamble. The damp proof course is measured across the opening and is adjusted later in the Example.

The wall in which the arch and opening occurs is measured as a blank wall previously. Items under the heading "Arch & Opg" adjust that previously measured by adding the work in the arch, etc., and deducting the work in the opening and that which is replaced by the arch.

Waste calculations take half the span of the arch as the height of a triangle and the arch radius as the hypotenuse and using trigonometry tables or a suitable calculator find the angle of the sector. The cosine of this angle is used with the radius to calculate the height to the springing of the arch. This deducted from the radius gives the rise of the arch (for use later). The arch is 328 mm high on face. Half this height added to the radius gives the mean radius of the arch.

Length of arc = number of degrees x radius x .017453. This formula is used to calculate the mean girth on face of the arch and also to calculate the length of the intrados.

The brickwork in the arch does not exceed one metre in thickness and is measured by area stating the thickness. The dimensions of the arch are the mean girth x the height. The area measured for the centering of the arch is that of the temporarily supported intrados.

The additional description given to describe the type of arch and the centering is thought to be helpful. It is not essential where there is but one type of arch which would be clearly identified by the use of standard descriptive features.

234 | BRICKWORK WALL   DRG No. U/D/1

Surf. features a.b.

       2325
        150
       2175

Jambs + soffit

2/ 0.33   {Fair facing   (U378
    2.18   (jambs
    2.83   (soffit
    0.33

                    Opg.
              Engin. bkw. a.b.
     2.72   {Ddt. Mass bkw., thickn
     2.33   {328 mm, vert walls, a.b.
2/3/ 2.72   (U331.1
     0.34
     2.99   (arch
     0.33
              Surf. features a.b.
2/   2.72   {Ddt. Fair facing
     2.18   (U378
2/2/3/ 2.72
     0.34

              Bkw. ancills.
     2.72   Ddt. Pitch polymer
              d.p.c. Width 328mm
                    (U382.2

(4)

COMMENTARY

The dimensions attached to the adjoining description "fair facing" are for the jambs of the opening and the soffit of the arch.

Brickwork in vertical walls is deducted for the opening and also where it is replaced by the brickwork of the arch. The dimensions for the segmental part of the opening are expressed in terms of the approximate formula chord x two thirds height of segment.

Fair facing and damp proof courses having previously been measured over the opening are deducted in the penultimate and last items, respectively.

17 Painting, Waterproofing and Miscellaneous Work—CESMM Classes: V, W and X

The rules of measurement for painting (Class V) and waterproofing (Class W) are to a great extent similar. The scope of miscellaneous work (Class X) is limited and does not merit a separate chapter. The three Classes are here dealt with in a single chapter.

PAINTING – CESMM CLASS V

Class V of the CESMM is applicable to in situ painting and in situ surface preparation which is carried out after the delivery of the components to the Site. Painting and other surface treatment carried out prior to delivery of components will be included with the components. For "Includes" and "Excludes" refer to the head of the Class V classification table in the CESMM.

In situ Painting (refer to Table 17.01)

Item descriptions for painting state the materials to be used and either the number of coats or the film thickness. They also describe the surfaces to be painted in accordance with tabulated Second Division features. Additionally, some items are further classified by appropriate Third Division features. (See Table 17.01, which gives also the units of measurement).

The First Division classifies paint types. Separate descriptive features are provided for primer paints. Items for painting systems which require surfaces to be primed in situ before the application of further painting treatment will give the priming separately when it is of a different type of paint from that used for the further painting treatment.

Separate items are not required for the preparation of surfaces. It being implied that the preparation is part of the painting. Consequently, it is necessary for the painting items to identify the preparation where it would not otherwise be clear.

Painting Surfaces Other than Metal Sections and Pipework

Plate VW.1 illustrates angles of inclination for painting surfaces exceeding one metre in width, other than surfaces of metal sections and pipework.

Painting surfaces not exceeding one metre in width (other than those of metal sections and pipework) are measured linearly stating tabulated width ranges. The Third Division descriptive features use the word "width" (not girth) which on strict interpretation means each surface width is classified as a surface of appropriate width. For example, the classification of each of the four faces of a component of small square cross-section as a surface of appropriate width. It is found more practical to classify the surfaces of such members according to their developed width, i.e. the aggregate of the four connected widths and to note in Preamble that this convention has been adopted.

The margin between the lower and upper limits of the standard width classifications, for painting measured linearly, is fairly large. There will be cases where it will be of benefit to classify some surface widths with greater precision, or to indicate location.

Table 17.01   In situ Painting

Generally - No deductions for holes and openings each less than 0.5 m2 (Note V3). Separate items not required for preparation (Note V7)	

1st Division	
Lead, iron or zinc based primer paint	State the materials to be used and either the number of coats or the film thickness (Note V1)
Etch primer paint	
Oil paint	
Alkyd gloss paint	
Emulsion paint	
Cement paint	
Epoxy or polyurethane paint	
Bituminous or coal tar paint	

2nd Division - Surface		3rd Division	
Metal, other than metal sections and pipework		Classify in accordance with the tabulated 3rd Division ranges:-	
Timber		(i) surfaces exceeding 1 m wide, by angle of inclination (Plate VW.1.)	m2
Smooth concrete			
Rough concrete		(ii) surfaces not exceeding 1 m wide, by width. Not distinguished by inclination (Note V2)	m
Masonry			
Brickwork		Use the classification "isolated groups of surfaces" where more convenient to number rather than measure in detail. The classification is applicable where total area of a group does not exceed 6 m2. When this classification is used, descriptions must identify the work to be painted and state its location. Groups of same shape and dimension may be included in one item (Note V4).	nr
Metal sections	Ignore additional area of connecting plates, brackets, rivets, bolts, nuts and similar projections (Note V5)		m2
Pipework	Measure length over valves and fittings and multiply by barrel girth of the pipework with no addition for flanges, valves or projecting fittings (Note V6)		m2

COMMENTARY

Painting Metal Sections

The area measured for painting metal sections is the overall length multiplied by the girth of the members, ignoring any additional area of the projections of connecting plates, brackets, rivets, bolts and the like.

COMMENTARY

In situ Painting (cont.

Painting Metal Sections (cont.

The dimensions which would be taken for the painting of the steel column shown in Figure V1 (assuming the size of section to be 305 x 305 mm) are given to the right of the diagram.

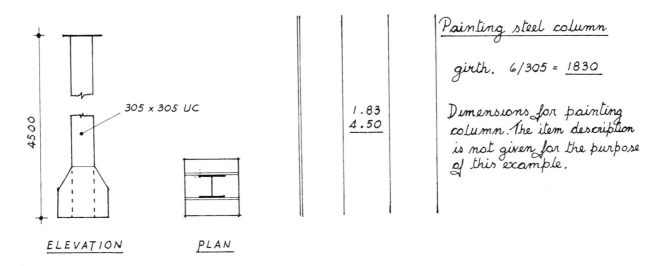

Fig. V1. Painting Steel Column

Painting Pipework

The area measured for painting pipework is the length, measured over valves and fittings, multiplied by the barrel girth of the pipework. The additional area of flanges, valves and other projecting fittings is not measured. The barrel girth for un-lagged pipework will be that of the pipe. The barrel girth of lagged pipework will be that of the lagging. Painting pipework with flanges, painting pipework without flanges and painting lagged pipework, are each given separately.

An example of the dimensions which would be taken for painting the pipework shown in Figure V2, are given below to the right of the diagram

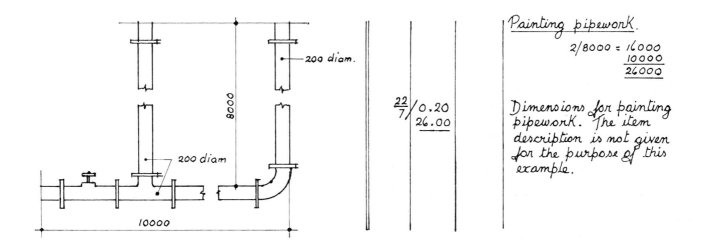

Fig. V2. Painting Pipework.

COMMENTARY

Specimen Item Descriptions for Painting

Number	Item description	Unit	COMMENTARY
	PAINTING		Note V1 and V7 of the CESMM make it necessary to amplify the standard descriptive features. In compliance with Note V1, the First Division features in the specimen items are amplified by reference to specification clauses where the type of paint would be fully described. The further requirement of Note V1 is met by stating the number of coats in the item descriptions.
V111	<u>Calcium plumbate primer paint to BS 3698, Type A, on metal other than metal sections and pipework</u> Galvanised steel upper surfaces inclined at an angle not exceeding 30 degrees to the horizontal; in one coat.	m2	
V170.1	<u>Red lead primer paint to BS 2523, Type B</u> Surfaces of previously primed steel sections; in one coat.	m2	It is necessary for the item descriptions for painting to identify the surface to which the paint is to be applied, so that its preparation may be allowed for in the painting items. Different cost characteristics arise between the preparation of ferrous surfaces and non-ferrous surfaces and also between treated, primed and un-primed surfaces. The specimen item descriptions, therefore, distinguish each particular type of surface.
V170.2	Surfaces of steel sections; in two coats.	m2	
V311	<u>Oil based paint as Specification clause VS 17.3, in three coats (two undercoats, one coat hard gloss finish), on previously primed metal other than metal sections and pipework</u> Steel upper surfaces inclined at an angle not exceeding 30 degrees to the horizontal.	m2	In the specimen item descriptions, the repetitive descriptive features form the subject of sub-headings and the variable features are listed under the appropriate sub-headings. This format may be adapted by putting less description in the sub-heading and more in the items which follow, or vice versa, according to personal preference or the requirements of a particular Bill.
V317	Steel surfaces of width 300 – 400 mm.	m	
V318	Steel isolated groups of surfaces; screen, outlet chamber.	nr	
V563	<u>P.V.A. based emulsion paint as Specification clause VS 17.7, in three coats, satin finish</u> Brickwork surfaces inclined at an angle exceeding 60 degrees to the horizontal.	m2	The width range stated in specimen item, Code V317, is defined to closer limits than the standard width range to allow of easier identification and to provide a sounder basis for adjustment of variations.
V880	<u>Pitch coating, as Specification clause 17.23, minimum dry film thickness 100 microns, on shop primed steel</u> Surfaces of pipework; flanged.	m2	

PAINTING CESMM, CLASS:V	ZONES of inclination relative to features in 3rd Division	WATERPROOFING CESMM, CLASS:W
3rd Division Upper surfaces inclined at an angle not exceeding 30 degrees to the horizontal		3rd Division Upper surfaces inclined at an angle of not exceeding 30 degrees to the horizontal
Upper surfaces inclined at 30 - 60 degrees to the horizontal		Upper surfaces inclined at 30 - 60 degrees to the horizontal
Surfaces inclined at an angle exceeding 60 degrees to the horizontal		Surfaces inclined at an angle exceeding 60 degrees to the horizontal
Soffit surfaces and lower surfaces inclined at an angle not exceeding 60 degrees to the horizontal		Not applicable to Waterproofing

PLATE VW.1.

Specific "Includes" in Class W of the CESMM are damp-proofing, tanking and roofing. Waterproofed joints and damp-proof courses are excluded from the Class, as noted in the "Excludes" at the head of the classification table in the CESMM.

Table 17.02 Waterproofing

Generally – State the materials to be used and the number and the thickness of coatings or layers (Note W1)

Measure area as that of the surface covered (Note W4)

No deduction for holes and openings each less than 0.5 m2 (Note W4)

Separate items not required for preparing surfaces or for joints, overlaps, mitres, angles, fillets and built-up edges or for laying to falls or cambers (Note W2)

1st Division	2nd Division	3rd Division	
Damp-proofing Tanking Roofing	Asphalt Sheet metal of stated material Waterproof sheeting Waterproof coatings Rendering in ordin-ary cement mortar Rendering in water-proof cement mortar	Upper surfaces not exceeding 30 degrees to the horizontal (See Plate VW.1.) Upper surfaces 30 – 60 degrees to the horizontal (See Plate VW.1.) Surfaces exceeding 60 degrees to the horizontal (See Plate VW.1.) Curved surfaces (less than 10 m radius (Note W5) Domed surfaces (less than 10 m radius (Note W5)	m2 m2 m2 m2 m2
Protective layers	Sand asphalt Flexible sheeting Sand Sand and cement screed Tiles	Surfaces of a width not exceeding 300 mm (not to be distinguished by inclination or curvature Note W3) Surfaces of width 300 mm – 1 m (not to be distinguished by inclination or curvature Note W3) Isolated groups of surfaces (area of group not to exceed 6 m2. Groups of same shape and dimension may be given in one item Note W6)	m m nr
Sprayed or brushed waterproofing m2			

COMMENTARY

Waterproofing (refer to Table 17.02)

Features together with a summary of CESMM Notes and units of measurement for waterproofing are set out in Table 17.02. A comparison of this Table with Table 17.01 indicates that the classification "soffit surfaces" is provided for painting but not for waterproofing. The classification "curved surfaces" and "domed surfaces" are provided for waterproofing but not for painting. Apart from these differences the rules for measuring waterproofing are similar to those for measuring painting.

Item descriptions describe the work in accordance with the descriptive features listed in Table 17.02. Second division features are amplified to state the materials and the number of coatings or layers. At Third Division level, features provide for work to surfaces one metre wide and greater, other than that to surfaces curved or domed to less than 10 m radius, to be described according to listed angles of inclination (See Plate VW1). The descriptive features "curved surfaces" and "domed surfaces" are applicable to work applied to such surfaces which are less than 10 m radius and one metre in width or greater. Standard features for work to surfaces of a width not exceeding one metre provide for width to be described as either "width not exceeding 300 mm" or "width 300 mm - 1 m" as appropriate. Item descriptions for surfaces of a width not exceeding one metre do not distinguish inclination or curvature. Work to surfaces not exceeding one metre in width is given in linear metres. That to surfaces one metre wide and greater is given in m2. See subsequent Commentary for work described as to "isolated groups of surfaces".

Isolated Groups of Surfaces

Waterproofing to a group of surfaces of a total surface area not exceeding 6 m2, such as that to a sump, an isolated plinth or the like, may be given as a numbered item rather than as a series of items measured as previously described. Item descriptions for the numbered items identify the work to be waterproofed and state location (See notes against the 3rd Division feature on Table 17.02).

Separate Items Not Required

Requirements will be specified for the work for which separate items are not required (See notes in first panel of Table 17.02). Tenderers need to ascertain the extent of the work from the drawings and allow for it in the items for the measured work or elsewhere in their tender. In respect of preparation, it is necessary, where the same waterproofing is applied to more than one base material and where there is more than one preparation requirement for the same base material, for the item description for the waterproofing to distinguish different preparation requirements.

Layers of subsidiary materials forming part of a waterproofing system are usually included with the main material in an item for the system. For example, isolating membranes and insulating board layers associated with asphalt would be described and included in items with the asphalt coatings. Non-structural screeds beneath waterproofing may also be included in such items, but where because of thickness differences their inclusion would unduly increase the extent of itemisation for the waterproofing, the screeds would be billed as separate items.

COMMENTARY

Waterproofing (cont.

Specimen Item Descriptions

Specimen item descriptions for the asphalt tanking to the floor and walls of the basement shown in Figure W1 which gives a part of the cross-section, are given at the side of the diagram.

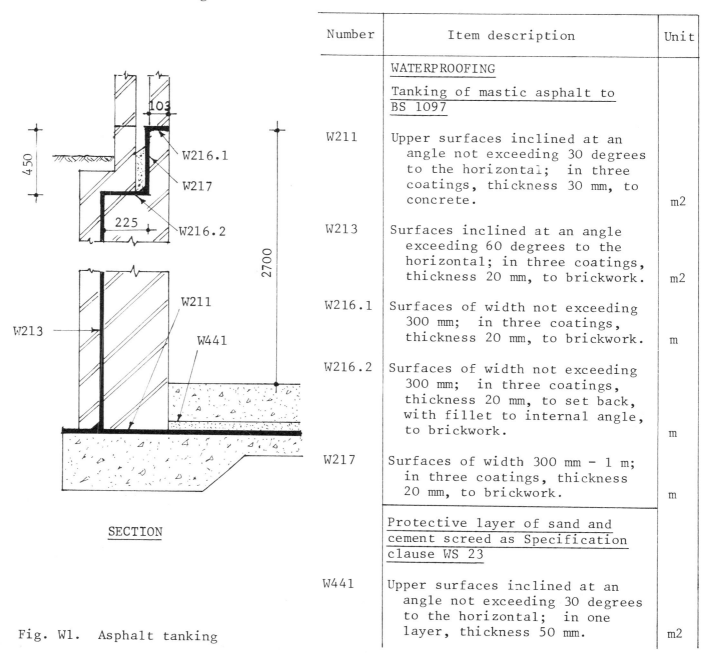

Number	Item description	Unit
	WATERPROOFING Tanking of mastic asphalt to BS 1097	
W211	Upper surfaces inclined at an angle not exceeding 30 degrees to the horizontal; in three coatings, thickness 30 mm, to concrete.	m2
W213	Surfaces inclined at an angle exceeding 60 degrees to the horizontal; in three coatings, thickness 20 mm, to brickwork.	m2
W216.1	Surfaces of width not exceeding 300 mm; in three coatings, thickness 20 mm, to brickwork.	m
W216.2	Surfaces of width not exceeding 300 mm; in three coatings, thickness 20 mm, to set back, with fillet to internal angle, to brickwork.	m
W217	Surfaces of width 300 mm – 1 m; in three coatings, thickness 20 mm, to brickwork.	m
	Protective layer of sand and cement screed as Specification clause WS 23	
W441	Upper surfaces inclined at an angle not exceeding 30 degrees to the horizontal; in one layer, thickness 50 mm.	m2

SECTION

Fig. W1. Asphalt tanking

Where it is considered it will simplify pricing, or aid identification, waterproofing work not exceeding one metre in width may be given in items which state actual width in preference to grouping several widths in an item which uses the standard width range given in the Class W classification table of the CESMM.

Table 17.03 Fences, Gates and Stiles and Drainage to Structures above ground

1st Division		2nd Division	3rd Division	
Fences	State types and principal dimensions of both fences and foundations (Note X1) State when erected on a curve of radius not exceeding 100 m or on a surface inclined at an angle exceeding 10 degrees (Note X2)	Classify in accordance with 2nd Division features	Measure height from Commencing Surface (Note X3)	m
Gates and Stiles	State type and principal dimensions of component and also those of foundations (Note X1)	Classify in accordance with 2nd Division features	Measure width between inside faces of posts (Note X3)	nr
Drainage to Structures above ground	State types, principal dimensions and materials of components (Note X4)	Classify in accordance with 2nd Division features	Gutters Fittings to gutters Downpipes Fittings to downpipes Identify fittings in accordance with Note X5	m nr m nr

COMMENTARY

Fences, Gates and Stiles (refer to Table 17.03)

The "Includes" and "Excludes" listed in Class X and Class E, respectively, of the CESMM, infers that excavation and foundations are deemed to be included in the Class X items for fences, gates and stiles.

Class X provides for fencing to be given as an all inclusive linear item. When billing safety fences it is often more convenient to enumerate end anchorages, additional intermediate posts and such like, rather than separately itemising various lengths to distinguish differences. Where this is done amending Preamble is necessary indicating that notwithstanding that the CESMM makes no provision for their separate itemisation, the stated items have been given separately in the Bill of Quantities.

Drainage to Structures above ground (refer to Table 17.03)

In the absence of any guidance in Class X of the CESMM, the rule in Note 5 of Class I, is applied and pipes and gutters are measured along their centre lines inclusive of the lengths occupied by fittings. Additional enumerated items, which identify the types of fittings, are given for any fittings.

Where substantial or unusual brackets or supports are used for gutters or downpipes, they are usually given separately from the linear items. Otherwise, the linear items for gutters are usually given to include the fixing brackets or holderbats.

COMMENTARY

Specimen Item Descriptions for Miscellaneous Work

Number	Item description	Unit
	MISCELLANEOUS WORK	
	Refer to Preamble clause 17.1	
	Fences, as Specification clause 4.3	
X112	Four railed, timber post and rail stockproofed fence, height 1.20 m above finished surface of verges, driven posts, as Drawing No. 4D/4.	m
	Safety fences, as Specification clause 4.8	
X191	Single sided tensioned rail safety fences, height 750 mm above surface of hard shoulder, as Drawing No. 4D/6.	m
	Gates, as Specification clause 4.5	
X216	Oak single field gates, width 4.25 m, height 1.30 m, with painted mild steel hangings and fastenings, 200 x 200 mm oak posts and 450 x 450 x 600 mm grade 20 concrete foundations.	nr
	Drainage to structures above ground	
	Cast iron gutters	
X321	Box gutters, width 200 mm, depth 150 mm, on brackets to brickwork.	m
X322.1	Stop ends on 200 x 150 mm gutters.	nr
X322.2	Outlets for 150 mm diameter pipes on 200 x 150 mm gutters.	nr
	Cast iron pipes	
X323	Downpipes, diameter 150 mm, with holderbats to brickwork.	m
X324	Anti-splash shoes on 150 mm diameter downpipes.	nr

COMMENTARY

The Commencing Surface for fencing may be difficult to determine. It seems practical to measure the height from a stated finished surface of the Works. If this is done amending Preamble is required. In this Example it is given the hypothetical clause number 17.1. Wording in the Bill of Quantities would be on the following lines:-

Preamble clause 17.1

Heights of fences are measured from the surfaces stated in the item descriptions, notwithstanding that Note X3 of the CESMM states they shall be measured from the Commencing Surface.

Note X1 of the CESMM calls for principal dimensions to be given in the item descriptions. Item Code X216 in the Example illustrates a description giving the detail. An abbreviated description giving a Drawing or Specification reference, which gives the same detail will often be substituted.

It is assumed that the gutters and downpipes are clearly identified by the descriptions given in the Example and that a Drawing or Specification reference is not necessary for this purpose.

Although billed here with fencing and gates, it is likely that the gutters and downpipes would be given in the Part of the Bill for the structure or building to which they relate.

Index